Information Security and Cryptography

Series Editors

David Basin, Department of Computer Science, ETH Zürich, Zürich, Switzerland

Kenny Paterson, Department of Computer Science, ETH Zürich, Zürich, Switzerland

Editorial Board

Michael Backes, Department of Computer Science, Saarland University, Saarbrücken, Germany

Gilles Barthe, IMDEA Software Institute, Pozuelo de Alarcón, Spain

Ronald Cramer, CWI, Amsterdam, The Netherlands

Ivan Damgård, Department of Computer Science, Aarhus University, Aarhus, Denmark

Robert H. Deng, School of Information Systems, Singapore Management University, Singapore, Singapore

Christopher Kruegel, Department of Computer Science, University of California, Santa Barbara, Santa Barbara, USA

Tatsuaki Okamoto, Okamoto Research Laboratory, NTT Secure Platform Laboratories, Musashino-shi, Japan

Adrian Perrig, CAB F 85.1, ETH Zurich, Zürich, Switzerland

Bart Preneel, Department Elektrotechniek-ESAT /COSIC, University of Leuven, Leuven, Belgium

Carmela Troncoso, Security and Privacy Engineering Lab, École Polytechnique Fédérale de Lausa, Lausanne, Switzerland

Moti Yung, Google Inc, New York, USA

Kui Ren, University at Buffalo, Buffalo, USA

Information Security – protecting information in potentially hostile environments – is a crucial factor in the growth of information-based processes in industry, business, and administration. Cryptography is a key technology for achieving information security in communications, computer systems, electronic commerce, and in the emerging information society.

Springer's Information Security & Cryptography (IS&C) book series covers all relevant topics, ranging from theory to advanced applications. The intended audience includes students, researchers and practitioners.

Reynaldo Gil-Pons · Ross Horne · Sjouke Mauw ·
Felix Stutz · Semen Yurkov

Security Protocols and Threat Models

Security and Privacy via The Applied
π-Calculus

Reynaldo Gil-Pons
Department of Computer Science
University of Luxembourg
Esch-sur-Alzette, Luxembourg

Sjouke Mauw
Department of Computer Science
University of Luxembourg
Esch-sur-Alzette, Luxembourg

Semen Yurkov
Department of Computer Science
University of Luxembourg
Esch-sur-Alzette, Luxembourg

Ross Horne
Department of Computer and Information Sciences
University of Strathclyde
Glasgow, UK

Felix Stutz
Department of Computer Science
University of Luxembourg
Esch-sur-Alzette, Luxembourg

ISSN 1619-7100 ISSN 2197-845X (electronic)
Information Security and Cryptography
ISBN 978-3-032-08248-0 ISBN 978-3-032-08249-7 (eBook)
https://doi.org/10.1007/978-3-032-08249-7

© The Editor(s) (if applicable) and The Author(s), under exclusive license to Springer Nature Switzerland AG 2026

This work is subject to copyright. All rights are solely and exclusively licensed by the Publisher, whether the whole or part of the material is concerned, specifically the rights of translation, reprinting, reuse of illustrations, recitation, broadcasting, reproduction on microfilms or in any other physical way, and transmission or information storage and retrieval, electronic adaptation, computer software, or by similar or dissimilar methodology now known or hereafter developed.
The use of general descriptive names, registered names, trademarks, service marks, etc. in this publication does not imply, even in the absence of a specific statement, that such names are exempt from the relevant protective laws and regulations and therefore free for general use.
The publisher, the authors and the editors are safe to assume that the advice and information in this book are believed to be true and accurate at the date of publication. Neither the publisher nor the authors or the editors give a warranty, expressed or implied, with respect to the material contained herein or for any errors or omissions that may have been made. The publisher remains neutral with regard to jurisdictional claims in published maps and institutional affiliations.

This Springer imprint is published by the registered company Springer Nature Switzerland AG
The registered company address is: Gewerbestrasse 11, 6330 Cham, Switzerland

If disposing of this product, please recycle the paper.

Preface

This book supports a range of reader profiles, having been used to support graduate courses and also to introduce researchers and security professionals to the field of security protocols. The book provides a compact theory of security protocols, along with a philosophy for how to use the theory for their design and security analysis. The primary aim is to present a theory that can be used to understand a range of approaches to security encompassing model checkers, proof assistants and pen-and-paper methods.

The theory presented is based on a state-of-the-art presentation of the applied π-calculus – a calculus defining how networks behave in the presence of threats. However, we start with simpler building blocks and plenty of intuition, with the aim to make the material accessible to graduate students and security professionals who may not yet have encountered such calculi. The book begins by addressing how to model threats to protocols used in networked systems that can lead to problems including confidential secrets being revealed or attackers hijacking the identity of honest agents. We explain how advanced threats and authentication problems, such as relay resistance, are confounded by the use of wireless technology. The book also covers strong privacy properties of protocols, their verification and methodologies for describing attacks whenever they exist.

This book can be understood without knowing how the underlying cryptographic schemes are implemented, e.g., how stream ciphers are constructed, the group theory underlying signature schemes, or how key pairs are generated. There are many excellent texts covering cryptographic schemes such as the classic texts of van Oorchot [88] and of Katz and Lindell [75]. There are other resources closer to the methodology presented in this book, such as the following. There is a text that we point readers to for further information on the tool Proverif by Blanchet [28]. The manual of the tool ProVerif [31] is regularly maintained and is particularly useful as a supporting reference for the syntax of the tool. There is a tutorial providing a complementary view on the theory of symbolic protocol analysis by Cortier and Kremer [45]. In the same series as this book, on Information Security and Cryptography, there is also a book by Basin, Cremers, Dreier and Sasse that provides a related approach to protocol verification focusing on using the tool Tamarin [87, 22]. We recommend these text as additional authoritative sources particularly if additional examples of protocols are required while revising the material, having completed the exercises provided. The current book however is however self contained and develops transferable skills, facilitating logical thinking about security protocols and the adversarial thinking behind security in general, making the book an ideal gateway to the literature for career-long learning.

This book was originally developed to provide lecture notes, labs, and tutorials for graduate courses in cyber security and in information security. The course has been developed at both University of Strathclyde, Glasgow, and University of Luxembourg. It has been used as the basis of master courses of various duration including 14 weeks, 10 weeks, and 7 weeks, with 2 to 4 hours contact time per week and around 150 hours of private study. The book has also been used for PhD student training. The initial chapters, which can be followed without the full technical apparatus of later chapters, are suitable for an introduction to security protocols within a broader module in the areas of information security and cyber security.

Undergraduate proficiency in discrete mathematics (set theory, logic, automata, etc.) is assumed as a prerequisite. For readers wishing to refresh themselves, there are numerous textbooks on discrete mathematics at various levels that may be skimmed through [60, 27]. For readers looking deeper into the theory, or seeking a simpler process calculus to gain familiarity with before tackling the applied π-calculus and its applications to security protocols, we recommend the books on the π-calculus by Robin Milner [91] and Sangiorgi & Walker [103]. These do not cover the applied π-calculus, but can help with understanding the underlying theory of concurrent communicating processes which this book applies to security.

Code for exercises can be downloaded from Springer Nature Code Inside: `https://github.com/sn-code-inside/security-protocols-and-threat-models`.

Glasgow and Luxembourg

Reynaldo Gil-Pons
Ross Horne
Sjouke Mauw
Felix Stutz
Semen Yurkov

Contents

Part I Introduction and Intuition

1 Introduction 3
   1.1 Protocols are everywhere 3
   1.2 Vulnerabilities discovered in payment protocols 4
   1.3 Fundamentals of security protocols: using this book 5
   1.4 Ethics: the responsible disclosure of vulnerabilities 7

2 Motivating case studies 9
   2.1 Introduction 9
   2.2 Motivating case study 1: The design of an ePassport protocol 9
      2.2.1 Security and privacy requirements 10
      2.2.2 Pitfall 1: Non-injective vs. injective authentication 11
      2.2.3 Pitfall 2: Unilateral authentication vs. mutual authentication 15
      2.2.4 Pitfall 3: Authentication vs. secrecy 16
      2.2.5 Pitfall 4: Reflection attacks 18
      2.2.6 Tool support 19
      2.2.7 Practice for getting started informally 22
      2.2.8 Some computational considerations: The length of secrets 23
      2.2.9 Further considerations: message integrity 25
      2.2.10 Practice in modelling nonce exposure 26
      2.2.11 Limitations of the BAC protocol for ePassports 27
   2.3 Motivating case study 2: Threats to the EMV protocol 28
      2.3.1 EMV standard overview 29
      2.3.2 Initialisation 32
      2.3.3 Offline Data Authentication 34
      2.3.4 Cardholder Verification 36
      2.3.5 Transaction Authorisation 38
      2.3.6 Examples of insecure EMV configurations 41

Part II Modelling Protocols and Threats

3 Syntax and operational semantics for security protocols 47
   3.1 Introduction 47
   3.2 Symbolic reasoning on syntax in computer science 48
        3.2.1 A syntax for messages and threads 49
        3.2.2 Preliminaries on substitutions 52
        3.2.3 Practice with the concepts of abstract syntax and substitutions 56
        3.2.4 Modelling threads in a Message Sequence Chart 57
        3.2.5 Practice in modelling protocol roles as threads 59
        3.2.6 A digression into music and back to security protocols 60
   3.3 Modelling protocol behaviour using operational semantics 61
        3.3.1 An equational approach to defining functions on messages .. 62
        3.3.2 States for labelled transitions.......................... 63
        3.3.3 The labelled transition system: inference rules 64
        3.3.4 A formulation of secrecy and traces 74
        3.3.5 Practice in operational semantics for threads.............. 76
   3.4 From threads to networks 79
        3.4.1 An operational semantics for networks 80
        3.4.2 Practice with the operational semantics for networks 84

4 Threat models for open networks 85
   4.1 Introduction 85
   4.2 Why the enterprise architecture assumption is outdated 86
   4.3 Modelling realistic networks such as the internet 89
        4.3.1 Other tools such as Scyther 96
        4.3.2 Practice in verifying secrecy in open networks 97
   4.4 Discovering attacks with the help of proof systems 98
        4.4.1 A sequent calculus for messages 99
        4.4.2 Examples of a deduction 100
        4.4.3 Practice using the sequent calculus..................... 101
   4.5 Calculating input labels using the sequent calculus 101
        4.5.1 An attack on the Dolev-Yao 2 cascade protocol:
                a fully worked example 103
        4.5.2 Practice in discovering attacks using sequents 116
   4.6 Presenting proofs that secrecy properties hold 118
        4.6.1 Practice in proving secrecy 121
        4.6.2 Practice in advanced aspects of semantics 122

Part III Authentication and Privacy

5 Authentication by agreement 127
   5.1 Introduction 127
   5.2 Agreement – a strong authentication property 128
        5.2.1 Example attack on agreement 130
        5.2.2 Agreement in ProVerif................................ 134
        5.2.3 Attack on agreement requiring an open network 136

		5.2.4	Attack on injectivity	138
		5.2.5	Practice in verifying agreement	140
		5.2.6	Open networks and symmetric key cryptography	140
	5.3	Multiparty case study: OpenID Connect		142
		5.3.1	Authorization Code Grant protocol	143
		5.3.2	An identity provider mixup attack	145
		5.3.3	Practice in multiparty protocols and ProVerif	146
		5.3.4	Practice in multiparty agreement	150

6 Authentication using time and distance ... 157
6.1 Introduction ... 157
6.2 A rudimentary authentication property: recent aliveness ... 157
6.2.1 Assessing the extent of aliveness vulnerabilities ... 160
6.2.2 Practice in verifying weak aliveness ... 166
6.2.3 Practice on recent aliveness involving XOR ... 167
6.3 Resisting relay attacks using distance-bounding protocols ... 168
6.3.1 Mafia fraud: attacks subverted by distance bounding ... 171
6.3.2 Distance hijacking: when the attacker is far away ... 177
6.3.3 Practice in verifying distance-bounding protocols ... 181
6.4 Tooling for aliveness and distance bounding ... 184
6.4.1 Tooling for aliveness ... 184
6.4.2 Practice in Tooling for Mafia fraud ... 187

7 Privacy ... 191
7.1 Introduction ... 191
7.2 The strong privacy property: unlinkability ... 192
7.2.1 Bisimilarity: checking there is no distinguishing strategy ... 195
7.2.2 Hennessy-Milner and the modal logic classical $\mathcal{FM}$ ... 198
7.2.3 Privacy in card payments ... 199
7.2.4 Practice in verifying privacy ... 202
7.3 Putting it all together ... 204
7.3.1 Recapping agreement and distance bounding ... 205
7.3.2 Unlinkability in ProVerif ... 210
7.3.3 Forward secrecy and beyond ... 211
7.3.4 Practice in using tools on larger case studies ... 213

References ... 217

Index ... 223

List of Figures

2.1   A protocol that does not satisfy injective authentication. 13
2.2   Injectively authenticating, but not mutually authenticating. 15
2.3   Mutually authenticating, but does not establish a shared secret. 17
2.4   Vulnerable to a reflection attacks or mutually authenticating. 18
2.5   Towards the ICAO 9303 standard for ePassports. 24
2.6   A representation of the ICAO 9303 standard for ePassports. 25
2.7   Payment architecture. 29
2.8   The EMV protocol stages. 30
2.9   Initialisation of the EMV protocol. 32
2.10   The Static Data Authentication mode. 34
2.11   The Dynamic Data Authentication mode. 35
2.12   CVM: Offline cleartext PIN. 37
2.13   CVM: Offline encrypted PIN mode. 38
2.14   Offline transaction authorisation. 39
2.15   Online transaction authorisation. 40
2.16   Long-known vulnerability of the EMV protocol 42
2.17   Overview of the strategy of an attacker. 43
2.18   Attack bypassing the PIN in an offline transaction. 44

3.1   An abstract syntax tree for a message. 49
3.2   Comparing the ProVerif, applied π-calculus and Scyther syntax. 51
3.3   Graphical representation of binders and bound variables. 53
3.4   Dolev-Yao 1: a broken unilateral secret-sharing protocol. 58
3.5   The Needham-Schroeder protocol. 60
3.6   Rules of a labelled transition system for threads. 66
3.7   The derivations for the three transitions of our running example. 72
3.8   Extending the labelled transition system rules to networks. 80
3.9   An example involving REP-ACT. 83

4.1   Informal representation of attack on Dolev-Yao 1 using an MSC. ... 92
4.2   The attack on Dolev-Yao 1 from Fig. 4.1, expressed formally. 92

4.3	Dolev-Yao 2: a cascade protocol, as introduced by Dolev and Yao	97
4.4	Dolev-Yao 3: a corrected version of the protocol in Fig. 4.3	98
4.5	Rules of the sequent calculus for our equational theory E.	99
4.6	Labelled transition system annotated with constraint systems.	107
4.7	Attack on the Dolev-Yao 2 protocol.	117
4.8	A specification of the Dolev-Yao 3 protocol.	119
5.1	Agreement-Example protocol.	131
5.2	An attack on agreement for Agreement-Example, MSC representation.	133
5.3	The Yo protocol.	137
5.4	Attack on non-injective synchronisation for the Yo protocol.	138
5.5	Unilateral authentication.	139
5.6	Protocols for the exercises on agreement.	140
5.7	A protocol based on the Authorisation Code Grant flow.	144
5.8	Attack on the secrecy of the code from the perspective of the browser.	146
5.9	A thread modelling the App from Fig. 5.7.	154
5.10	A thread modelling the Browser from Fig. 5.7.	155
5.11	A thread modelling the Issuer from Fig. 5.7.	156
6.1	EMV configuration featuring Static Data Authentication ODA.	158
6.2	Attack on recent aliveness of EMV with SDA.	159
6.3	Basic protocols failing weak aliveness.	161
6.4	A protocol satisfying weak aliveness in the correct roles.	164
6.5	Protocol for aliveness (Exercise 6.1).	166
6.6	Protocol involving XOR.	167
6.7	General structure of a distance-bounding protocol.	169
6.8	Meadows 1 protocol.	170
6.9	The attacker is nearby and the honest prover is far away.	172
6.10	TREAD Protocol (simplified, public key version).	173
6.11	Timing scenarios for distance-bounding protocols.	175
6.12	Mafia fraud attack on TREAD Public Key Protocol.	176
6.13	Attacker is far and only honest verifiers and provers are close.	177
6.14	Meadows 2 Protocol.	178
6.15	A distance-hijacking attack.	179
6.16	An example distance-bounding protocol for Exercise 6.3.	182
6.17	Another example distance-bounding protocol Exercise 6.3.	182
6.18	An example distance-bounding protocol for Exercise 6.4.	183
7.1	Message sequence chart for BAC as in the French ePassport.	194
7.2	Attack on French ePassports.	197
7.3	Blinded Diffie-Hellman: handshake for EMV 2nd gen protocol.	199
7.4	The Kim-Choi-Lee protocol.	203
7.5	Distance-bounding key agreement for card payments that satisfies unlinkability, injective agreement and secrecy of the session key.	206
7.6	The PACE ePassport protocol.	214

Part I
Introduction and Intuition

Chapter 1
Introduction

Abstract The importance of security protocols is explained firstly by pointing out their ubiquity. Known authentication vulnerabilities in payment systems are used to illustrate current pressing security issues, and hence the need for a methodological approach to address such vulnerabilities. Various learning curves for reading subsequent chapters are suggested, with core exercises and deeper study paths highlighted. Ethical considerations, notably responsible disclosure, are underscored for all readers aiming to apply such security knowledge to real protocols.

1.1 Protocols are everywhere

Security protocols are used in critical infrastructure, such as banks, military networks, mobile phone networks, and airports. They are at the foundations of technologies that impose on our every-day lives and enable our increasingly smarter cities. They protect our confidential information, such as financial accounts and business secrets; but, today this is not enough. Security protocols should also protect our identities, ideally preventing our online presence from being easily detected and our habits and activities tracked by parties with dubious intentions. It should be a concern that, in practice, many security and privacy properties are not preserved by protocols, even when they are stated objectives.

A wealth of attacks have been discovered on widely-deployed critical infrastructure by using symbolic methods, some of which we will highlight in this introduction and elaborated on further in this book. Typically, symbolic approaches seek to discover logical flaws in how a security protocol is assembled, based on a model of how a protocol is executed in the presence of threat actors. Once mastered, such methods can be applied by security professionals and researchers to methodologically check for critical design flaws, or substantiate security arguments. The emergence of powerful tools and fundamental advances demand this book on the topic.

1.2 Vulnerabilities discovered in payment protocols

We begin our story in 2025, when we are aware of a host of related vulnerabilities in contactless payments [94, 50, 11, 25, 24, 99, 42]. Whenever any of us performs a contactless payment using a credit or debit card, the card talks to the payment terminal at the point of sale using the EMV protocol. The EMV protocol was designed by Europay, Mastercard and Visa to ensure that all terminals could read all cards, and is now used by almost all banks, worldwide. Some of the vulnerabilities discovered allow an attacker to exploit a card, possibly in your pocket, to make a payment for a large sum of money. We will cover two fundamental shortcomings of the EMV protocol in this book.

The first problem is that anyone with a wireless device, e.g., a smartphone, can start a conversation with a card and relay the messages to a point of sale without consent of the card holder. This kind of relay attack, by itself, is not the biggest problem, since, without a PIN, the card can only make small payments of, say, maximum 100 euros or thereabout. In contrast to activating the card which will engage in a session with any device that is sufficiently close, the entering of a PIN which is not compromised requires a conscious physical effort from the cardholder. Therefore, an argument that banks might use to defend this well-known vulnerability is that, since the financial reward is low for the attacker and there is a risk of being caught and prosecuted, a simple relay is not an attack strategy that they believe will be exploited frequently enough to impact insurance payouts.

The second problem is more serious. An attacker can fool the terminal into believing that the payment was verified, not by the user entering a PIN, but, instead, via another means such as device which can be used to make payments such as a smartphone. This is an *authentication* vulnerability. In this context that means that, although the terminal is physically engaged in a conversation with the card, at the end of running the protocol the terminal is tricked into believing that it was in conversation with another device such as a smartphone that appears to have already approved the payment. Consequently, since the terminal believes the payment was authorised by a smartphone, it does not ask for a PIN and hence may be prepared to approve payments worth thousands of euros. In short, the attack strategy is to exploit the fact that there are multiple modes of payment and to fool the card and terminal into engaging in different modes resulting in critical checks being skipped.

In more general terms, authentication is the idea that, if an agent (e.g., a point-of-sale terminal) executes a protocol with an agent they trust (e.g., a card issued and certified by the bank in the above case), then one agent can trust that another agent really did execute the protocol as expected. For example, if a card is present, the steps were taken to approve the payment by the card in full according to the correct protocol for verifying that type of card. There are many variants of authentication where the precise expected behaviour of communication partners differs. In the case of the vulnerability described above the precise property violated is known to security experts as *injective agreement*, which is the expectation that the parties (e.g., the card) created a new session of the protocol for each successful authentication attempt and every message sent and received was really the message received by the authenticating party (e.g., the terminal). If that property holds, then no message could be manipulated in transit without causing the protocol to abort. We will cover injective agreement in this book.

The authentication flaw in the EMV protocol should be of concern to the public, since billions of payment cards are in circulation worldwide. It is likely that you, the reader, hold such a card, in which case you yourself may be vulnerable to attacks that could be used to defraud you of thousands of euros just by carrying your contactless card in public, depending on the implementation of the card.

From the above story concerning vulnerabilities in the EMV protocol, we can gain insight into the role of symbolic verification methods in the discovery and anlysis of such attacks. Without going through the process of turning the specification into a formal model, perhaps we would not know that on a particular line in one of around a dozen documents that specify the EMV protocol there is something written that causes authentication to be broken in a very particular sense. The process of modelling the protocol was, in itself, a key step in discovering some of these vulnerabilities systematically, where modelling protocols is a process that requires expertise that we aim to foster in this book. Indeed, going through the systematic security thinking in the process of modelling itself can be the way in which an expert spots flaws. The verification tools which then check models of protocols are then useful for backing up proposed fixes, since not just known attacks are explored by such tools but also other potential vulnerabilities are explored that could be exploited via unforeseen attack strategies. Given that protocols such as EMV are widely deployed, we can reasonably ask whether symbolic verification should be conducted when such a security protocol is designed rather than retrospectively.

1.3 Fundamentals of security protocols: using this book

Authentication, as discussed in the story above, can be subtle to capture. Secrecy is more obviously the absence of the ability for an attacker to learn a secret message or key once it is agreed upon. Secrecy is important for confidentiality, preventing attackers from learning session keys that allow them to eavesdrop on a secure channel for instance. In the early days, security protocols were deployed where even secrecy was not preserved. The most famous example is the Needham-Schroeder protocol [95], which was shown to be completely broken when used on a modern network, such as the internet, where everyone can connect to everyone [81]. Expertise in protocol design has matured, thereby reducing the incidence of vulnerabilities violating secrecy. However, mistakes still can be made, when developers design their own protocols or assemble them from components.

The first core message we wish to underscore is that, even if you use the best post-quantum cryptography, if you use it when assembling a protocol in the wrong way, your protocol can still be broken. Our second core message is that problems leading to secrecy being compromised can be counterintuitive, making them beyond the scope of engineering intuition alone. For example, we will study examples where it is possible to take a good protocol and break it by adding extra layers of encryption, which may seem counterintuitive. To illustrate these two points first, we begin this book by focussing on the security property of secrecy in Chapters 3 and 4, before proceeding with more subtle properties in Chapters 5, 6 and 7 that are less immediately obvious and hence more likely to require scrutiny in the design of modern protocols.

The book is structured to allow different modes of study. Chapter 2 provides motivating scenarios that aim to build an intuition for security protocols and security properties through examples representative of their real world usage. That chapter can be read without a formal understanding of symbolic methods and can be safely skipped by readers wishing to get straight to the formal content. The chapter has been used to support motivating lectures and introductory lab sessions where tools can be set up and demoed, via core exercises Ex. 2.1, 2.2, and 2.3 and advanced exercise Ex. 2.4.

Chapters 3 to 7 are supported by exercises some of which we suggest as core, in the sense that completing them supports the learning curve moving forward. The proposed core exercises are Ex. 3.1, 3.2, 3.3, 4.1, 4.2, 5.1, 6.1, 6.3, and 7.1. Further exercises are provided throughout the book for a deeper study path, which may be skipped while obtaining sufficient understanding of the material to solve realistic problems. Such questions are for readers aiming for greater mastery of the methods or, perhaps, to support revision on a second reading. Those more advanced exercises are Ex. 3.5, 4.3, 4.4, 4.5, 5.2, 5.3, 6.2, 6.4, 6.5, 7.2, 7.3, and 7.4. Experience when teaching using this material suggests that there is a steeper learning curve for Chapters 3 and 4 than other parts of the book, but getting over that curve by studying the core material makes the remaining topics readily accessible. Chapters 5, 6 and 7 can be sampled from independently in any order to gain expertise in authentication and privacy.

What may be tricky to understand at first is the operational semantics – the model of the execution of protocols – which is a rich nontrivial piece of machinery. The operational semantics is carefully designed, incorporating ideas from several decades of research, to cleanly separate the internal state of honest agents from the capabilities of attackers infiltrating a network. Much of the explanation of the operational semantics is isolated in Chapter 3. A suggested strategy is to read that chapter but not to get hung up on the operational semantics, and instead to move forward and try to apply the resulting models to the security problems in subsequent chapters. This suggestion is to emphasise that the primary purpose of this book is on honing skills that can solve security problems. The operational semantics is a tool that helps readers form a mental model that they can use to systematically approach security protocol properties, and which can also be useful for understanding the inputs and outputs of tools that can support the verification of such properties.

1.4 Ethics: the responsible disclosure of vulnerabilities

There is an ethical dimension to research in security in general, including research into security protocols. If you study systems that are really used, such as the EMV protocol which is part of the critical infrastructure enabling us to make ePayments, then you risk being successful. It is possible that from time to time you uncover vulnerabilities in protocols or related aspects of the system which, when perhaps combined with some social engineering, can bypass security features, exposing the system to attacks that are unanticipated by the engineers that designed the protocol.

Thus, before engaging in research on security, it is essential that we are aware of ethical processes that are used to mitigate legal and moral risks associated with publishing vulnerabilities. There is no official ethical process, however a widely accepted guideline would be to inform stakeholders, such as EMVCo, Europay, Mastercard and Visa, the relevant national ministry, or other organisations responsible for the technology in which your vulnerability was discovered. This can be done by email or mail; however, it is best to execute this process using lawyers, who will assist in formulating legal aspects of a letter to better protect your rights. Rights to preserve include the retention of your intellectual property and your right to publish your results in the public domain.

After informing the relevant stakeholders, it is best practice to give the stakeholders time to respond. For instance, a stakeholder may respond to request that a serious vulnerability is not published in public domain. Good practice would be to leave at least 90 days for responses before the vulnerability is published. This provides time for the relevant stakeholders to review your description of the vulnerability, distribute internally to relevant committees, and prepare their own private response, perhaps to yourself, the press, and other stakeholders such as banks and the general public.

After 90 days, without a response, or with a response that does not raise objections to publication, it is typically acceptable to publish your results. In your publication process, you must take care to precisely and transparently describe the vulnerability discovered. Indeed, it is your ethical responsibility that, if you are aware of a vulnerability, then you must make the vulnerability known to others in a timely fashion. The conventional wisdom is that "security by obscurity" is never a solution. That means that it is better that a vulnerability is published to facilitate timely mitigation strategies by those affected and to avoid a situation where malicious attackers independently discover and exploit related vulnerabilities.

For a recent example, in September 2019, some of the authors of this book published details on a privacy vulnerability they discovered in the ICAO 9303 standard for ePassports [59]. Before publishing this paper, we informed ICAO, the UN agency responsible for ePassports, via a responsible disclosure process similar to that described above. This provided ample time for ICAO to contact the standard body ISO and coordinate a joint statement when the vulnerability was made public. This discussion is in the public domain, since it was reported on in detail by the press and raised in the parliament of Luxembourg [68]. Due to the ethical process that we followed we are able to discuss the vulnerability that we reported to ICAO in this book. For the same reason, we may discuss the vulnerabilities communicated by Toro et al. in the EMV protocol; or other vulnerabilities such as those discovered in authentication protocols use in the 5G networks that are intended to secure our conversations [23].

The authors also have experience disclosing vulnerabilities in the open-source projects and community-driven standards. While the experience is quite different, the process is roughly the same. The relevant chairs and members of the community were contacted indicating that there is a vulnerability. In such community-driven open-source project probably do not assume that the people contacted are aware of the ethics process described here or importance of the security problem, thus we suggest being kind to such communities. Join their channels and observe whether they respond and politely nudge them if the disclosure is not passed to and examined by the appropriate members of the community. The way in which these more open communities are structured means that you may become part of their solution if you can explain clearly the threat and your evaluation of countermeasures.

It is important to accept ethical standards before conducting security research, or even embarking on this book. We, as security researchers, do not advocate the approach of hackers such as the one who stole $600 million of cryptocurrency and then returned it to expose a vulnerability in smart contracts. This is aptly explained in an article covering this particular incident [106], explaining that it was wrong to label that hacker as a "white hat hacker" – a term intended for hackers that responsibly uncover vulnerabilities in order to improve systems. Instead, the vulnerability discovered in smart contracts should have been responsibly disclosed within the framework of the law. This interdisciplinary problem of connecting cyber-security research and policy is a challenge in itself requiring better understanding between legal experts and security experts, and also clarification of the relevant laws and professional ethical codes. There may also be regional laws and guidelines that apply, particularly if systems involved impact critical infrastructure or the handling of corporate or state secrets. An advantage of the methodology supported by this book in this respect is that we need not touch a real system or handle personal data to analyse a protocol, since we verify a model of the protocol rather than testing a real system directly.

Chapter 2
Motivating case studies

Abstract This chapter serves to form an intuition about security protocols and their security properties, without requiring formal understanding. Two case studies are presented. The first case study develops an ePassport protocol by starting with a simple protocol that is not secure and iteratively improving it such that successively stronger secrecy and authentication properties are satisfied. Exercises provide an introductory hands-on tutorial, giving an early taste of using tools to verify those security protocols. The second case study examines compelling authentication vulnerabilities in the EMV protocol used for making payments.

2.1 Introduction

In this chapter, we form an intuition for some key security and privacy properties of protocols that are defined precisely in later chapters of this book. Often equipping ourselves with an informal intuition first makes it easier to then go on to formally understand security protocols and make precise judgements about security claims. By taking a look into real-world protocols we also gain a concrete feel for the ubiquitous role of security protocols in our everyday lives. Aspects of these case studies will be revisited more formally later in this book. The two case studies cover the BAC ePassport protocol and the EMV protocol used for making payments using a smartcard or phone.

2.2 Motivating case study 1: The design of an ePassport protocol

In this section, we explore informally the design of a real security protocol for ePassports. Most countries now issue electronic passports, or ePassports, that implement a protocol standardised by the ICAO – the International Civil Aviation Organization, which is a UN agency. By standardising the protocol by which a reader communicates with an ePassport, an ePassport from any country can be read by a reader at a checkpoint in any other country. Anyone can operate an ePassport reader, for example there are apps available for NFC enabled smartphones.[1]

Here, we explore the design of one of the original protocols for ePassports, by examining the requirements of such an ePassport protocol. We will present a series of naïve protocols and explain their strengths and security flaws in order to understand the design decisions made.

2.2.1 Security and privacy requirements

Later in this book, we will formally and precisely define security properties of protocols, so we can make precise judgements and make use of tools for determining whether or not security properties hold. Initially, for this section, we present only a very informal idea of key properties, since the precise definitions are rather involved, making them difficult to grasp initially without forming an intuition.

For the authentication protocol used in ePassports, a cryptographic key is derived from information printed on the biometric page of the ePassport which is loaded into the Reader. Usually this information for deriving the key is transferred by an OCR (Optical Character Recognition) session when you present the ePassport to a scanner; although this information is printed and may also be entered manually. The key is then used by both the ePassport and the Reader in an exchange of wireless messages between the Reader and a chip embedded inside a page of the ePassport, with the aim to securely transmit personal and biometric information stored on the chip to the Reader. The protocol must satisfy several security properties, where confidentiality of the biometric information information is just one goal.

Security property 1: authentication. When the protocol is run successfully, the ePassport must know it is really communicating with the Reader which was presented with they key – the ePassport must authenticate the Reader. That is, the ePassport is really talking to the reader that was presented with the printed keys of the ePassport, as opposed to another device perhaps controlled by an attacker. Conversely, the Reader should also authenticate the ePassport, which means that when the Reader runs the protocol successfully, it should also be the case that the Reader really was communicating with the ePassport for which these keys were provided.

Security property 2: secrecy. After having passed authentication, the ePassport and reader should have established a shared session key that will be used to transmit encrypted personal data between the ePassport and reader for that session only. We should know that the key cannot be intercepted or reconstructed in any way by an attacker. This is an example of a secrecy property.

These two properties are typical core requirements for an authentication protocol, although other properties may also be desirable as we will touch on later.

---

[1] An example app available through app stores at the time of writing is the VISOGO app maintained by INCERT in Luxembourg: https://www.incert.lu/solutions/#visogo

2.2 Motivating case study 1: The design of an ePassport protocol   11

The threat model. For both of the above properties we should be clear about the threat model, which defines the capabilities of an attacker that attempts to violate the security property. The intention here is to model an attacker that is capable of placing wireless devices controlled by the attacker in the vicinity of a reader when it is interacting with an ePassport. To do so, we make the following assumptions.

1. We make the strong assumption that an ePassport shares its key with trusted Readers only who do not abuse their knowledge of the key, nor are they compromised devices. This is usually achieved via OCR and so we assume that an attacker operating wireless devices does not have the capability to intercept the printed key obtained in an OCR session with such a Reader.
2. An attacker can eavesdrop on radio-frequency communications between the Reader and ePassport. The attacker can therefore read all messages exchanged between the ePassport and Reader. This is a passive capability, that can be performed without interfering with the communication itself. This kind of eavesdropping is possible at a range of several meters.
3. An attacker can jam messages and transmit new messages. This includes initiating sessions using other devices, but without knowing the key of the ePassport. It has been demonstrated that active attacks can be perpetuated wirelessly at a distance of up to 1 meter, where the main obstacle is the need to provide power to the chip in the ePassport, which does not have its own power source [63]. The attacker may also have access to more intimate covert devices such as 180 micron thick overlay cards for intercepting and manipulating messages [12].
4. An attacker can use another network such as a mobile or other wireless network to share messages between devices. This gives the attacker full control of the wireless network, since it can simultaneously be in proximity to multiple devices and share information between the locations.
5. An attacker cannot – in reasonable time – decrypt messages without a key, reverse cryptographic hash functions, or guess keys and cryptographically generated random numbers. Such fresh random numbers that are generated in a session to be used once, are referred to in security protocols as *nonces*. The assumption is that nonces are sufficiently random that they cannot be easily guessed.

The above roughly corresponds to a very standard class of threat model known as the *Dolev-Yao threat model*. Attackers in the Dolev-Yao model can intercept the messages *output* by honest agents in a protocol, such as the genuine ePassport and reader, and use those messages to construct new messages which they can induce as *inputs* for the honest agents. There are other threat models with different capabilities, but we focus on Dolev-Yao threats throughout this book.

2.2.2 Pitfall 1: Non-injective vs. injective authentication

We start with a very simple design for a (fictional) protocol and iteratively improve it until we obtain a real protocol. Our first attempt at designing an ePassport protocol appears in Fig. 2.1 where it is depicted as a message sequence chart [84]. A message sequence chart (MSC) is a graphical representation of a protocol, which makes use of syntactic representations of cryptographic messages. We explain the elements of an MSC next by reference to Fig. 2.1:

- The two vertical threads labelled with the roles *ePassport* and *Reader* represent the timeline of the two types of role that an honest agent may assume when executing a session of this protocol.
- The information ke above the ePassport represents a symmetric key, derived from information printed on the ePassport, that is to be used to encrypt data. It appears above the ePassport role, since initially it is known only by the ePassport.
- The directed dotted line, in this diagram represent an OCR session with the Reader whereby the Reader obtains the key. This is a message that we have assumed cannot be intercepted by an attacker.
- The box containing the keyword *fresh* indicates that the Reader generates a nonce called nr (the nonce of the Reader).
- The directed solid line represents a wireless communication where the Reader sends a message to the ePassport. This is a message that the attacker may intercept, and attempt to manipulate.
- The message transmitted is represented on the label of the solid line. For this protocol, the message consists of the nonce nr, generated by the Reader, appended to a constant message "hello", and then encrypted with the encryption key ke of the ePassport. This is represented syntactically by the term $\{(nr, hello)\}_{ke}$. Notice this syntactic representation of a ciphertext abstracts away from the exact encryption mechanism used.
- The two boxes at the end of the Reader thread represent security properties from the perspective of the ePassport having completed the protocol. The first holds, but the second does not hold. We will discuss informally these properties below.

> **Note on message syntax: pairs and encryption.**
> You have already observed in the protocol in Fig. 2.1 that different message terms can be represented in syntax. Messages are either *atomic*, represented by variables, or *constructed* from functions. Atomic messages are typically represent private keys, constants, nonces, agent names, etc. Constructs are used to model cryptographic functions applied to them and represent, e.g., encrypted messages, hashes, public keys, bitstrings, etc. In this section, we will use the following constructions:
>
> - *Pair*: (M, N) represents a pair of two messages, M and N. Anyone who has a pair (M, N) can access its first and second elements using the first and second projection functions, satisfying the following equations: $\mathsf{fst}((M, N)) =_D M$ and $\mathsf{snd}((M, N)) =_D N$. It is straightforward to encode lists of arbitrary length using the pair operation. For instance, $(M, N, K) := (M, (N, K))$. Whenever it is clear from the context, e.g.,

2.2 Motivating case study 1: The design of an ePassport protocol

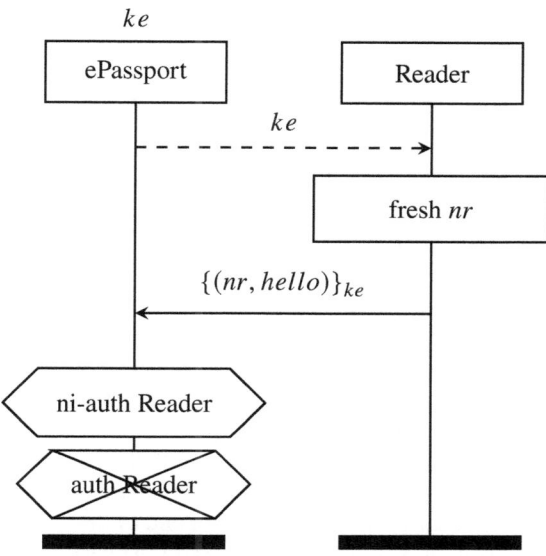

Fig. 2.1 A protocol that does not satisfy injective authentication.

when several messages are sent in one go or the pair is encrypted, we will drop parentheses representing pairs entirely.
- *Symmetric encryption*: $\{M\}_N$ represents the message M encrypted with the key N. To access the message M, one must have access to the key N and apply the decryption function dec, which satisfies $\text{dec}(\{M\}_N, N) =_D M$. As suggested above, we may drop parenthesis when encrypting pairs e.g., $\{(nr, hello)\}_{ke}$ can be represented as $\{nr, hello\}_{ke}$.

Functions like fst, snd, or dec that simplify a constructed term are called *destructors*, while functions ( ), h, or { } that "complicate" the term are called *constructors*.

Notice that we use a special symbol $=_D$ to describe the behaviour of destructors with respect to symmetric encryption. Equations such as the ones above, which explain how functions applied to messages interact, comprise the *equational theory* and are discussed formally in the next chapter.

We assume that, when executing a protocol, many instances of the protocol may be executed. Each time the same key ke is used, but a new nonce nr is generated. This reflects the fact that an ePassport will be willing to engage in a session with as many Readers as desired. This fact we will use to expose a vulnerability in the simple protocol in Fig. 2.1.

The protocol does appear to authenticate the Reader, from the perspective of the ePassport, and keep the nonce nr secret. To gain an impression why, observe that only an honest reader receives the encryption key ke and never the attacker. Therefore, the message $\{nr, hello\}_{ke}$ may only be created by the honest Reader or honest ePassport. The ePassport never transmits a message through the wireless communication channel (the solid line), so the message must originate from an honest Reader. The ePassport can verify this by decrypting the message using its own symmetric key and checking that the second part of the decrypted message is the constant "hello". Upon receiving and checking the message, the ePassport has confidence that: firstly, an honest Reader was really present; secondly, the message is exactly the message sent by the Reader, unmodified by an attacker; and, thirdly, the nonce nr must be secret, known only by the honest Reader and the ePassport. Note that this is not a formal proof, but we will develop the machinery for proving such properties later in this book.

The common weakness in the above argument, is that, although true, it neglects to recognise that there is a *replay attack*, which is a common attack pattern (producing an MSC describing the attack is an exercise at the end of this section). In particular, an attacker may replay a message sent by a reader in the past and the ePassport will not be able to detect that it is engaging with an honest Reader but in an outdated session that may have long-ago been terminated.

The above mentioned vulnerability is an attack on an injective form of authentication. Here injective means one-to-one in the sense that there is one ePassport session for every Reader session. Recall that an *injective* (or one-to-one) function is one that always maps two distinct elements to two distinct elements. By analogy, here, there is an injective function from events representing the completion of an instance of an ePassport running the protocol to events representing the initialisation of a Reader.

We should insist that a good authentication protocol is injective; yet we have just demonstrated that the protocol in Fig. 2.1 is not injective. After running the protocol once, the attacker knows the message $\{nr, hello\}_{ke}$ and can fool the ePassport into believing that it is authenticating a Reader as many times as it likes simply by replaying the message $\{nr, hello\}_{ke}$ that the attacker intercepted. This is despite the attacker not being able to decrypt the message and obtain the nonce nr.

For the above reasons, the security properties in Fig. 2.1 suggest that the ePassport does "non-injectively authenticate" the Reader, indicated by "ni-auth Reader". However, the ePassport does not fully authenticate the Reader, in the sense that there is not an injective mapping from ePassport sessions to Reader sessions, as indicated by the property "auth Reader" being crossed out.

2.2.3 Pitfall 2: Unilateral authentication vs. mutual authentication

Now, we introduce our second iteration of the protocol converging towards the design of a real protocol. Consider the protocol in Fig. 2.2, where, as in the previous protocol, the key of the ePassport is shared with an honest reader via an OCR session. Two messages are exchanged wirelessly: the first is a nonce nt generated freshly by the ePassport for each session (upon being powered up by the reader), while the second is a ciphertext assembled by the Reader, making use of the nonce received at the previous step from the ePassport.

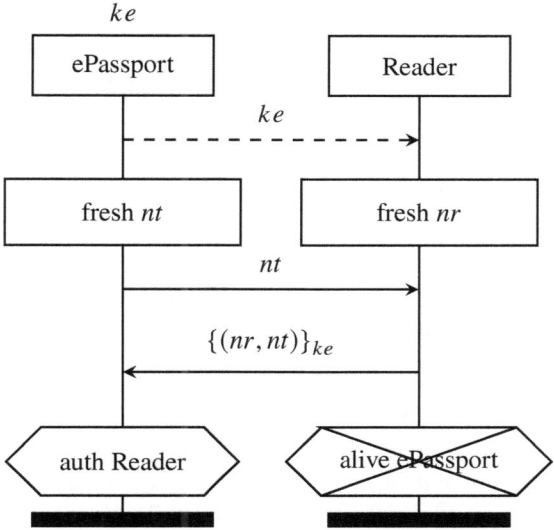

Fig. 2.2 A protocol that is injectively authenticating, but not mutually authenticating.

The protocol in Fig. 2.2 ensures that the ePassport authenticates the Reader, and does so injectively, avoiding the pitfall mentioned in the previous section. To see how the aforementioned replay attack is avoided, observe that for every new session of the protocol the ePassport generates a fresh nonce and that same nonce must be returned to the ePassport by the Reader for the ePassport to complete the protocol. There is no way for an attacker to fake that message, since it does not have the key ke. Therefore, the ePassport can uniquely match a message from a Reader to the protocol session corresponding to the nonce. This way, for every distinct execution of the ePassport protocol, there will be a distinct run of a Reader protocol that corresponds to it. In other words, the protocol is injective.

As noted previously, in this section, our arguments are not formal proofs, which require more machinery, developed later in this book. The aim of this section is to illustrate the kind of intuition we employ when we reason about security protocols. Indeed, a common problem is to miss an attack that is not immediately visible by applying experience only, and hence formal reasoning will be required to avoid protocols being deployed with such flaws.

The problem with the protocol in Fig. 2.2 is that, although the ePassport authenticates the Reader, the Reader does not authenticate the ePassport. Any good ePassport protocol should guarantee such *mutual authentication*, and hence this toy protocol falls short our requirements. This is easy to see, since the Reader only receives a nonce in cleartext, which anyone including the attacker could fake. Therefore, upon completing the protocol, there is no guarantee that any wireless messages were transmitted by the ePassport at all. We indicate this rudimentary failure of the Reader to authenticate the ePassport, by the crossed out property "alive ePassport" in Fig. 2.2. The failure of the passport to be necessarily "alive" from the perspective of the Reader indicates that there may be significant problems with the protocol.

2.2.4 Pitfall 3: Authentication vs. secrecy

Next, we present a protocol that does satisfy mutual authentication, that is both agents (injectively) authenticate each other. The protocol, depicted in Fig. 2.3 still does not meet all our requirements though, in a quite obvious sense: the protocol does not establish a shared secret.

We examine why the protocol is injectively authenticating, firstly from the perspective of the ePassport authenticating the Reader. From this perspective, the protocol is much the same as the protocol in the previous section, which, recall, is unilaterally injectively authenticating. In addition, the nonce nr is revealed to the attacker, after receiving the ciphertext $\{(nr, nt)\}_{ke}$. This release of nr is too late for the attacker to use that information to interfere with the first two messages. Thus, a similar argument to before applies for why the ePassport injectively authenticates the Reader.

Now consider why the protocol is injectively authenticating from the perspective of the Reader. Observe that the message $\{(nr, nt)\}_{ke}$, sent by the Reader, contains a nonce nr freshly generated by the Reader. This makes it impossible for an attacker to manufacture this message. Furthermore, since only the ePassport has the key ke, only the honest, genuine, ePassport can decrypt this message. Thus, if the nonce nr appears on the network, as in the final message of this protocol, then the Reader knows that the honest ePassport successfully read the message sent. Furthermore, since the nonce is different for each session, we can distinguish between each session, and hence the Reader will be able to injectively map each session it completes to a fresh session executed by the ePassport. Thus this protocol is a mutual authentication protocol.

Despite the protocol in Fig. 2.3 being mutually authenticating, secrecy however is trivially violated in the sense that there is no shared secret established. This is

2.2 Motivating case study 1: The design of an ePassport protocol

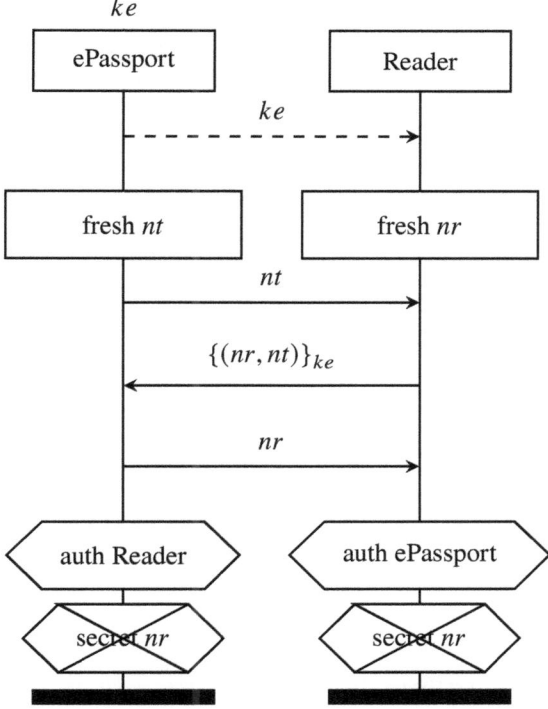

Fig. 2.3 A protocol that is mutually authenticating, but does not establish a shared secret.

immediate since both nonces are transmitted unencrypted, and hence can be intercepted easily by a passive attacker snooping on radio-frequency communications, even from a considerable distance given the right equipment.

Computational attacker models. Despite the above arguments, there are potential problems with authentication for this protocol that can be uncovered by a computational argument. The protocol can act as an oracle for a known plaintext attack, since the attacker learns a plaintext (nr, nt) and its corresponding ciphertext $\{(nr, nt)\}_{ke}$. The protocol can be run repeatedly to provide enough information to attempt a statistical attack that learns the key ke. Most symbolic threat models do not consider this kind of attack and consider it to be in the realm of computational attacks [9], which exploit imperfections in the encryption scheme employed or choice of keys. Computational threats are pertinent for ePassports, since the integrated circuits on such Radio-Frequency ID (RFID) tags have limited computational power so may not use strong enough schemes.

This view that the above mentioned attack is outside the scope of symbolic cryptography is not strictly true, since we can play with our assumptions, by including a rule in our attacker model to the following effect: if the attacker obtains a plaintext and ciphertext that match then they can, in worst case, obtain the key. This allows us to evaluate which security properties are broken as a result of obtaining that key. In the case of the protocol in Fig. 2.3, all properties would be broken in this stronger attacker model. This kind of playing with the attacker model is a central part of the philosophy of analysing security protocols – we are not done with security once we have established security in one threat model and so we should make our threat model explicit.

2.2.5 Pitfall 4: Reflection attacks

Consider the two protocols in Fig. 2.4. Both are variants of the protocol from the previous section where the last message is different: in one, the ePassport sends the message $\{(nr, nt)\}_{ke}$ back to the Reader, while in the other it sends $\{nr\}_{ke}$. Only one of these is mutually authenticating and establishes a shared secret.

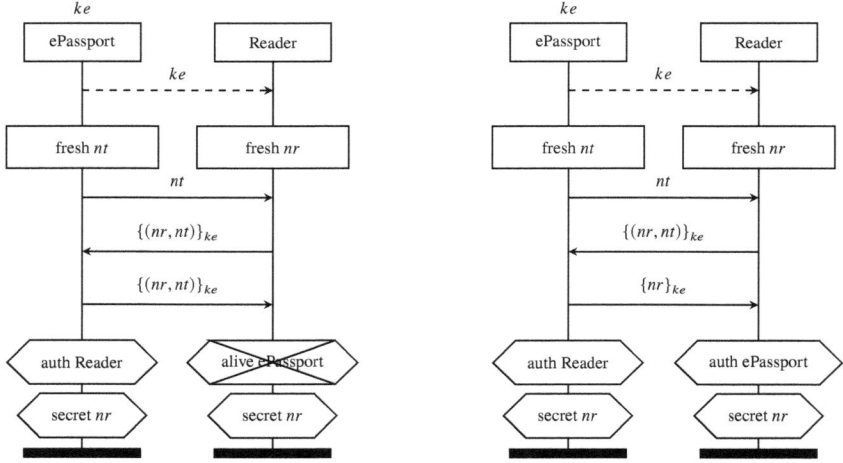

Fig. 2.4 Two protocols: the first is vulnerable to a reflection attack on authentication; while the latter is mutually authenticating.

The former protocol features what is known as a *reflection attack*. This is a specific type of replay attack, which can be exploited here by an attacker to fool a Reader into thinking that it has authenticated an ePassport without the ePassport having engaged in any (wireless) communications, and hence need not be alive. To

2.2 Motivating case study 1: The design of an ePassport protocol

see this, observe that an attacker can generate its own nonce, say *ne*, that it sends to the Reader. The Reader then responds with message $\{(nr, ne)\}_{ke}$, which the attacker learns. The attacker is therefore able to replay the same message $\{(nr, ne)\}_{ke}$ back to the Reader without the ePassport ever being present. That is, the message of the Reader is *reflected* back to itself. This breaks again the rudimentary aliveness property, as indicated in Fig. 2.4.

From this we learn that it is a good idea to ensure that two messages in the protocol should not be identical, even if there are symmetric aims, such as mutual authentication. A simple resolution to this problem is presented in the second protocol presented in Fig. 2.4. We leave an informal argument for why the second protocol is mutually authenticating and establishes a shared secret as an exercise.

2.2.6 Tool support

Tools can assist us with the analysis of security protocols, such as Scyther [47, 46], ProVerif [28] and Tamarin [87]. We will mainly focus on ProVerif since it is based on the applied π-calculus, to be introduced later, which is a powerful language for expressing how honest agents in a protocol interact with their untrusted environment.

We very briefly mention firstly the tool Scyther that is easy to use to get started.[2] The listing below shows the second protocol in Fig. 2.4 modelled in the protocol modelling language of Scyther.

protocol *example*(*T*, *R*) {
    role *T* {
        fresh *ni*: Nonce;
        var *nr*: Nonce;

        send_1(*T*, *R*, *ni*);
        recv_2(*R*, *T*, {*ni*, *nr*}k(*T*, *R*));
        send_3(*T*, *R*, {*nr*}k(*T*, *R*));

        claim_4(*T*, Nisynch);
        claim_5(*T*, secret *nr*);
    }
    role *R* {
        var *ni*: Nonce;
        fresh *nr*: Nonce;

        recv_1(*T*, *R*, *ni*);
        send_2(*R*, *T*, {*ni*, *nr*}k(*T*, *R*));
        recv_3(*T*, *R*, {*nr*}k(*T*, *R*));

---

[2] At the time of writing, the Scyther tool can be downloaded from the Web page of Cas Cremers: https://people.cispa.io/cas.cremers/scyther/

      claim_6(R, Nisynch);
      claim_7(T, secret nr);
  }
}

The two roles T and R represent an RFID Tag (e.g., an ePassport) and Reader respectively. The operations show the sequence of output and input that each role perform. To see how to move from the message sequence chart for a protocol to its representation in Scyther we perform a projection from the MSC to the *local* view of each role. To do this, informally, we firstly declare any fresh nonces generated by the role and variables representing the message received by each role. We then write down the series of messages that are input and output. Each such message is uniquely numbered and declares the sending and receiving roles of the message (e.g., R,T indicates a message from role R to role T), along with a representation of the message transmitted. The Scyther tool automatically verifies all the authentication and secrecy claims asserted.

In ProVerif, the same protocol and properties can also be represented. The code that we describe next can be executed using ProVerif.

> *Code Inside.* ProVerif can be obtained at `https://proverif.inria.fr`. To try out ProVerif quickly, `http://proverif16.paris.inria.fr` provides an online interface. Note $\wedge$ and $\implies$ in code listings abbreviate && and ==> in ProVerif syntax. The code explained here and in other chapters can be downloaded from Springer Nature Code Inside: `https://github.com/sn-code-inside/security-protocols-and-threat-models`.

Firstly, we need to provide ProVerif with an equational theory for symmetric encryption. This specifies that symmetric encryption allows a message encrypted with a key to be decrypted using the same key. Some type annotations help keep track of what represents a key and a bitstring.

type symkey.
fun senc(bitstring, symkey): bitstring.
reduc forall x: bitstring, y: symkey; sdec(senc(x, y), y) = x.

To specify secrecy and authentication properties in ProVerif, one defines events and queries. The events are used to indicate when certain things have happened and queries relate the events in order to express a security property. These queries formalise notions of secrecy and authentication as illustrated intuitively in this section.

The first query is false if a message m is believed to be secret, but is known to the attacker. We want this query to be false for secrecy to hold which explains why you will see that ProVerif automatically negates this query when it is executed.

2.2 Motivating case study 1: The design of an ePassport protocol

event confidential(bitstring). *(* secrecy claim *)*
query *m*: bitstring; **event**(confidential(*m*)) ∧ *attacker*(*m*).

For authentication, our query involves two events. The first event indicates when an agent believes it authenticated another agent and saw a tuple of messages. The second one indicates when an agent sent its last message and the tuples of messages that it saw. The query uses both events and is successful whenever there was a matching communication partner when an authentication claim is reached.

event auth(bitstring, bitstring, bitstring).
event sentLastOutput(bitstring, bitstring).
query a: bitstring, b: bitstring, *ms*: bitstring;
  inj-event(auth(a, b, *ms*)) ⟹ **inj-event**(sentLastOutput(b, *ms*)).

In ProVerif, we express the roles of the protocol in terms of inputs and outputs. The RFID tag is expressed as follows, where *t* and *r* are the identities of an RFID Tag and Reader respectively. Recall, we are modelling here the second protocol in Fig. 2.4.

let *Tag*(c: channel, *t*: bitstring, *ke*: symkey, *r*: bitstring) =
  new *n*: bitstring;
  out(c, *n*);
  in(c, *y*: bitstring);
  let (*n'*: bitstring, *m*: bitstring) = sdec(*y*, *ke*) **in**
  if *n* = *n'* **then**
  let *m2*: bitstring = senc(*m*, *ke*) **in**
  event sentLastOutput(*t*, (*n*, *y*, *m2*));
  out(c, *m2*);
  event confidential(*m*);
  event auth(*t*, *r*, (*n*, *y*)).

Notice the keywords **out** and **in** indicate inputs and outputs. When an input is received there are checks to determine whether the input is of the expected form. The events are related to the security properties expressed by the queries above.

Similarly, the Reader can be expressed by the following process.

let *Reader*(c: channel, *r*: bitstring, *ke*: symkey, *t*: bitstring) =
  in(c, *n*: bitstring);
  new *m*: bitstring;
  let *m2*: bitstring = senc((*n*, *m*), *ke*) **in**
  event sentLastOutput(*r*, (*n*, *m2*));
  out(c, *m2*);
  in(c, *z*: bitstring);
  let *m'*: bitstring = sdec(*z*, *ke*) **in**

if $m' = m$ **then**
event confidential(m);
event auth($r, t, (n, m2, z)$).

In ProVerif, we explicitly set up how the above processes talk to each other in a network.[3] The following is what is executed by ProVerif, when it is checking security properties.

free c: channel.
free a, b: bitstring.
process new *ke*: symkey; (
   (!*Tag*(c, a, *ke*, b)) |
   (!*Reader*(c, b, *ke*, a)))

The process above states that there are many copies of an RFID tag and a reader running in parallel and all sharing the same symmetric key. The key is known only by these two honest agents, as indicated by the binder **new**, while other information such as the identity of the tag and reader are free variables, and hence available to everyone.

2.2.7 Practice for getting started informally

The exercises below will help you gain a better informal understanding of authentication protocols.

Exercise 2.1 (Informal reasoning) The second protocol in Fig. 2.4 we stated is mutually (injectively) authenticating. Please provide an informal, but systematic explanation of why this is the case. You can make use of similar reasoning as in the subsections proceeding that figure. You must refer explicitly to the capabilities of the attacker, explaining what messages the attacker learns during execution and why they cannot break mutual authentication by influencing inputs.

Exercise 2.2 (Tools) With this book, we will learn how to better understand how to use protocol verification tools, what they do, and how to interpret their results. However, it is possible at this point already to view such tools more simplistically, as you might learn a programming language by example and trial and error. To give you an initial feel for these tools perform the following steps:

1. Download and install ProVerif or Scyther.
2. Enter the program provided above to reproduce the claimed results.

---

[3] Note that the syntax highlighting distinguishes between names that are bound by **free**, e.g., c, and the ones that are bound by **new**, e.g. *ke*.

2.2 Motivating case study 1: The design of an ePassport protocol   23

3. Modify this example, to obtain the protocols in Fig. 2.1, Fig. 2.2, Fig. 2.3, and Fig. 2.4. Note that the queries provided in the example above will be strong enough to pick up on all authentication vulnerabilities in this section, regardless of how they were named in the text. You may however remove or rename some of the events to guide the tool towards considering specific properties, e.g., a failure of secrecy by a specific agent only. The constant hello can be modelled by const hello:bitstring.
4. Summarise systematically in a table whether the attacks and proofs obtained by running the tool on these protocols match our informal assessment.

Exercise 2.3 (Cultural context) The Feldhofer protocol [56] is similar to the second protocol in Fig. 2.4. It was proposed as a minimal authentication protocol to respect the limited computational power of early RFID chips. The difference is that after receiving $\{(nr, nt)\}_{ke}$, the ePassport (or, more generally, an RFID tag) swaps the two messages, responding with $\{(nt, nr)\}_{ke}$.

1. Provide the message sequence chart for the Feldhofer protocol. You may refer to the original paper, cited above, which uses an alternative, but also very intuitive, notation to represent the protocol.
2. Explain what pitfall is avoided by swapping the two nonces in the final message.

2.2.8 Some computational considerations: The length of secrets

We almost have designed a real ePassport authentication protocol. Although the second protocol in Fig. 2.4 is authenticating and establishes a shared secret, the real Basic Authentication Protocol (BAC) used in the ePassports is slightly more involved. The BAC protocol is defined precisely by ICAO in Part 11 of the ICAO 9303 standard for Machine Readable Documents [71]. The protocol specification includes details such as how to derive the encryption keys from information on the biometric page of an ePassport, and details on encryption schemes.

The preamble to the BAC protocol specification states: "Exchanged nonces MUST be of size 8 bytes, exchanged keying material MUST be of size 16 bytes."

To understand the significance of this observation, consider the protocol in Fig. 2.5. This is very nearly the BAC protocol, and expands on the previous protocols as indicated by underlining. The key material exchanged is indicated as kr and kt. Observe that this data used for establishing a shared secret is different from the nonces nr and nt. The shared secret is then obtained by taking the XOR of the exchanged keying material, as indicated by $kr \oplus kt$.

Let us examine some advantages of this protocol design, where nonces and keying material are separated. We explain the importance of each of the four random numbers generated in the protocol. Firstly, 8 bytes is too short for the key material used to derive a session key since an 8-byte key is relatively easy to attack by brute-force compared to a 16-byte key. Since the private messages, containing sensitive private information such as biometrics will be transmitted using the session key, an

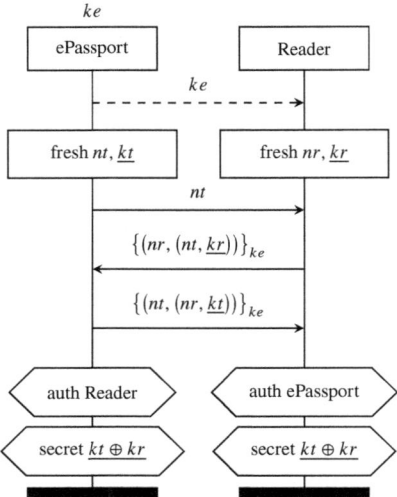

Fig. 2.5 Towards the ICAO 9303 standard for ePassports.

attacker may use brute-force to decrypt an intercepted encrypted session that follows if 8 bytes were used for the key material.

In contrast, 8 bytes is sufficient for the first nonce nt, since it is transmitted in cleartext, and therefore the risk is that an ePassport eventually reuses a nonce, and a message from an old honest Reader session is replayed to fool an ePassport into thinking it has authenticated an honest Reader that is not present. Firstly, the attacker would need to eavesdrop on or initiate sufficiently many sessions with an honest Readers loaded with the correct key, in order to have intercepted previously a ciphertext with exactly the same ePassport keys and nonce. Furthermore, it is even less likely for an attacker to then correctly guess the nonce in the given ciphertext that is replays when attempting to fake the response of the Reader.

The above suggests why nt is 8 bytes and kt is 16 bytes. Another reasonable question is: why is it insufficient to simply use kt from the ePassport as the key, use nr solely as part of authentication and omit kr from the protocol? A computational reason for including kr is that the RFID chip in the ePassport may not have the best dedicated circuit for cryptographic pseudorandom number generation, but a Reader is typically more powerful. Thus the randomness introduced by kr has higher entropy than kt, and hence $kr \oplus kt$ is of the quality of the device with the best cryptographic pseudo-random key generator. In addition, some authorities issuing passports might not trust how the cryptography in all Readers is engineered, even if the authority handling the Reader is trusted not to mishandle biometrics obtained via the session; so the converse argument can also hold, and hence kr alone cannot be used instead of kt. Thus both kr and kt appear in the protocol. Finally, observe that although nr is

2.2 Motivating case study 1: The design of an ePassport protocol

only 8 bytes and hence may be guessed in the future, the protocol is designed such that discovering the encrypted nonce *nr* in the future will not reveal any information about the sessions keys. Thus 8 bytes is arguably sufficient for *nr*, as used as a nonce in the protocol.

2.2.9 Further considerations: message integrity

As mentioned before, the ICAO [71] standard specifies the sequence of messages that form the BAC protocol. A fairly comprehensive representation of the sequence of messages that lead to successful authentication is summarised in Fig. 2.6, which extends the protocol previously presented.

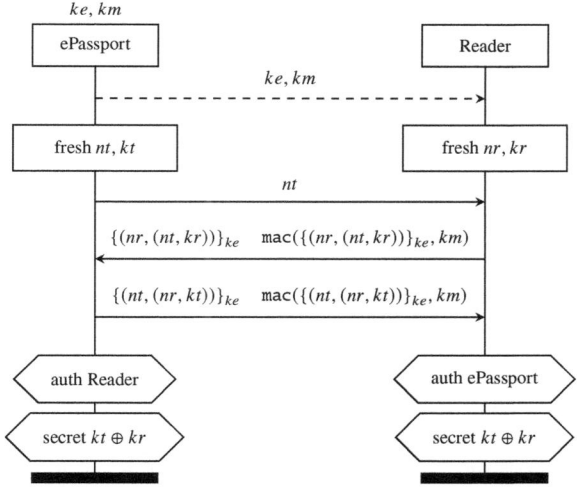

Fig. 2.6 A representation of the ICAO 9303 standard for ePassports.

Observe that the protocol is similar to the protocol we have defined so far except that a Message Authentication Code (MAC) is transmitted along with the message.

> **Note on message syntax: MAC operation.**
> The *MAC* operation is a form of cryptographic hash of a message M with respect to some secret that agents pre-share or derive, denoted $\mathrm{mac}(M, N)$. The secret N is often called the *seed*, and the value $\mathrm{mac}(M, N)$ the *MAC Value*. MACs are used to ensure *message integrity*, i.e., the ability to check that the message has not been modified in transit.

> To verify the integrity of the message M upon receiving $(M, \mathrm{mac}(M, N))$, one checks that the message M hashed with the secret N (which the receiving agent already has) coincides with the second element of the received pair $\mathrm{mac}(M, N)$. This way, if an attacker, for example, attempts to flip bits of a ciphertext with the objective of obtaining a new ciphertext for a related underlying message, the MAC value will pick up on the manipulation.

In this protocol, the key km, used as the seed of the MAC, is derived along with ke during the OCR session as indicated by the dotted line at the top of the protocol. By keeping this key secret from the attacker, only the honest agents in the protocol could have produced a MAC value corresponding to fresh message that is not identical to one that has already been transmitted.

An interesting property that MACs ensure for this protocol is that, even if the attacker were to have complete power to encrypt and decrypt messages, the Reader and ePassport still authenticate each other. Some encryption schemes have weaknesses such that they may be manipulated once part of the plaintext is known, e.g., observe that nt is known and appears in two different positions in encrypted messages in the protocol. Thus MACs can boost an encryption scheme that does not itself guarantee high message integrity.

2.2.10 Practice in modelling nonce exposure

The advanced exercise below explores the impact of nonce exposure on authentication and secrecy.

Exercise 2.4 (Optional) Model the effect that shorter 8-byte nonces cannot be guessed quickly enough to be used during the authentication of the protocol, but may be guessed after authentication has completed. Try modelling this by inserting an extra output revealing the nonce of the Reader at the end Reader process. How does this affect the analysis of the BAC protocol? Do the session keys remain secret, and does authentication still hold in your new threat model?

Now add MACs to your model and check that the intended properties still hold. To assist with getting started quickly, we provide the following definitions of roles.

let $Tag(c$: channel, t: bitstring, ke: symkey, km: symkey, r: bitstring$)$ =
  new n: bitstring;
  out(c, n);
  in$(c, (message2$: bitstring, $mac2$: bitstring$))$;
  let $(n'$: bitstring, m: bitstring$)$ = sdec$(message2, ke)$ **in**
  if $n = n'$ **then**
  if mac$(message2, km)$ = $mac2$ **then**
  let $message3$: bitstring = senc$((m, n), ke)$ **in**
  let $m3$: bitstring = $(message3$, mac$(message3, km))$ **in**

2.2 Motivating case study 1: The design of an ePassport protocol

event sentLastOutput(t, (n, ($message2, mac2$), $m3$));
out($c, m3$);
event confidential(m);
event auth(t, r, (n, ($message2, mac2$))).

let *Verifier*(c: channel, r: bitstring, ke: symkey, km: symkey, t: bitstring) =
in(c, n: bitstring);
new m: bitstring;
let *message2*: bitstring = senc((n, m), ke) **in**
let $m2$: bitstring = ($message2$, mac($message2, km$)) **in**
event sentLastOutput(r, ($n, m2$));
out($c, m2$);
in(c, ($message3$: bitstring, $mac3$: bitstring));
let (m': bitstring, n': bitstring) = sdec($message3, ke$) **in**
if $n' = n$ **then**
if $m' = m$ **then**
if mac($message3, km$) = $mac3$ **then**
event confidential(m);
event auth(r, t, ($n, m2$, ($message3, mac3$))).

Consider another threat model where the key for encryption is public, modelling that the encryption scheme is weak and hence messages may be manipulated by the attacker by some means. Assume however that the key for MACs is still secret. Check whether authentication still holds in that threat model, where *ke* is declared instead along with the free variables and only *km* is bound using new in the process.

2.2.11 Limitations of the BAC protocol for ePassports

Although reasonably well designed, as explored earlier, the BAC protocol does however have several known security and privacy limitations. We will return to some of these limitations later in this book, when exploring privacy, but summarise some weaknesses here briefly.

The BAC protocol has a serious limitation that the keys are generated as a function of data that includes information such as the name of the passport holder and their passport number. This means that the entropy is low, i.e., the ability to guess the key is high [80]. If the passport holder is known and the objective is to extract fingerprints for example, it is possible to reconstruct most of the key and therefore access in full the biometric data on the chip too easily. This is addressed in more recent ePassports that implement an improved authentication protocol in the ICAO 9303 standard called the Password Authenticated Connection Establishment (PACE) protocol [26], which uses a genuinely random PIN. Indeed, the EU recommends that ePassports of member states do not implement the BAC protocol in newly issued ePassports.

Another problem with the BAC protocol, also addressed by the PACE protocol, is that a security property called *forward secrecy* does not hold. Forward secrecy ensures that, if the long-term key of the ePassport is compromised in the future, then previously intercepted communications involving the ePassport remain secret [44]. Discovering a long-term key is a prevalent threat since, even without bruteforcing the key, an attacker may simply look at a scan of the biometric page in a hotel record for instance. Indeed, the energetic reader can already prove that BAC does not satisfy forward secrecy and the PACE protocol does satisfy this security property by using an extension of Scyther [21]. There are further privacy issues such as *unlinkability*, which is the inability of 3rd-party observers to track the movements of some ePassport holder [59]. This requires more sophisticated machinery to be analysed precisely, hence we delay discussion on this until the end of this book. In short, the BAC protocol is reasonably well designed, but has known imperfections.

2.3 Motivating case study 2: Threats to the EMV protocol

In the introduction, we mentioned the EMV protocol and some recently discovered attacks on authentication. Here, still informally, we expand on the EMV protocol, and explain some obvious known attacks in addition to the new attack, allowing to bypass PIN verification, thereby demonstrating that to this date EMV cannot be called secure. We also show, that EMV lacks privacy either, as the communication between the card and the terminal is not encrypted, allowing the identifying information about the card to be exposed during the transaction.

EMV is the most prevalent [51] payment method using smartcards – as of 2022 more than 91% of transactions are EMV. It came into place in the mid-1990s to replace magnetic stripe cards as they are incapable of computation and easy to clone. Payment providers such as American Express, Banrisul, Dankort, JCB, Mastercard, MIR, UnionPay, Visa, etc. follow the EMV standard [1, 2, 3, 4], a series of documents that specify how exactly payments should be done with the main focus on card-terminal communication. The standard is maintained by EMVCo, a consortium of payment processing companies. While there are other than EMV smartcard payment methods, e.g., parking lot cards or top-up shopping centre cards, their use is limited, and they follow proprietary protocols – in contrast, EMV is widely accepted and open. The standard is quite flexible: only minimal requirements must be respected, so it is up to the payment system that implements EMV, which additional options to include. Hence, the standard describes not a single protocol but a whole variety of configurations.

To start off, we introduce transaction flow and the respective infrastructure assumed by EMV in Fig. 2.7. The card C is manufactured by the *issuing bank BC* in collaboration with the payment system *PaySys* (e.g., Visa). The terminal T is connected to an *acquiring bank BT* supporting *PaySys* that processes payments on behalf of the terminal. The acquiring bank *BT* processes payments by connecting to the *PaySys* network that exchanges messages between banks.

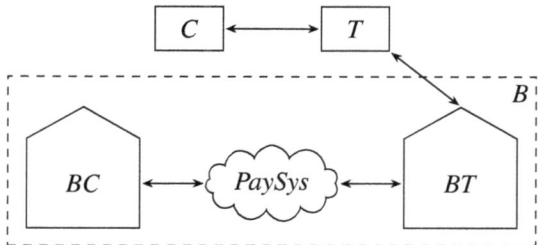

Fig. 2.7 Payment architecture.

A successful run of the protocol results in the generation of an Application Cryptogram *AC* by the card *C*. This Application Cryptogram *AC* is eventually sent by the terminal *T* to the acquiring bank *BT*, either before or after the payment is approved by the terminal, depending on whether the payment is online or offline, respectively. Upon receiving *AC*, the issuing bank *BC* decides to decline or accept the transaction, and replies with the appropriate message. The processing method on the banks' side, specified in the standard, is not mandatory as "issuers may decide to adopt other methods" [2]. In addition, the internal processing used by *PaySys* is proprietary and *PaySys*-specific. Therefore, when modelling the system, the agents *BT*, *PaySys*, and *BC* are often merged into a single agent *B*, modelling their common interface with the terminal when processing payments, as indicated in Fig. 2.7.

It was shown several times that some configurations allowed by the standard are not secure [49, 94, 25, 24, 99], making fraudulent transactions possible. The important aspect of EMV is that it also supports contactless cards, making it easier for a man-in-the-middle attacker to interact with the card without the cardholder being aware, thereby lowering the sophistication of, e.g., eavesdropping or relaying the communication between the card and the terminal. Passive eavesdropping, or active man-in-the-middle attacks, nonetheless, are still possible in the case of contact transactions – an unnoticeable "shim" properly installed inside the reader is enough for eavesdropping/collecting transaction data [33]; suitable devices that can alter messages and, for instance, trick the terminal that the right PIN was entered when the PIN is unknown, exist [94]. In the following chapters we systematically explore the EMV standard and highlight its potential vulnerabilities.

2.3.1 EMV standard overview

The full official specification of the EMV protocol is about 2000 pages long and is spread across multiple volumes. For the sake of completeness, we give a high-level description of a generic EMV transaction. We warn the reader that our goal is to demonstrate how convoluted the EMV standard is, and how easily an insecure configuration can be chosen for implementation, rather than giving a comprehensive description of all possible configurations of EMV. For the latter, we refer the reader to the technical report by van den Breekel et al. [37], where such description is given in a clear and concise form that still has about 33 pages.

We reiterate that the EMV standard mainly specifies communications between the card and the terminal. This communication consists of a series of command/response exchanges. Each command may come with a payload that either contains the expected data, as specified by the protocol, or provides data requested in a previous step. The latter case involves a so-called Data Object List (DOL), a list of data elements that must be sent in response. For instance, the Processing Options Data Object List (PDOL) is a special case of DOL that a card uses to request specific transaction data that the terminal should provide, e.g., currency, amount, country code, etc.

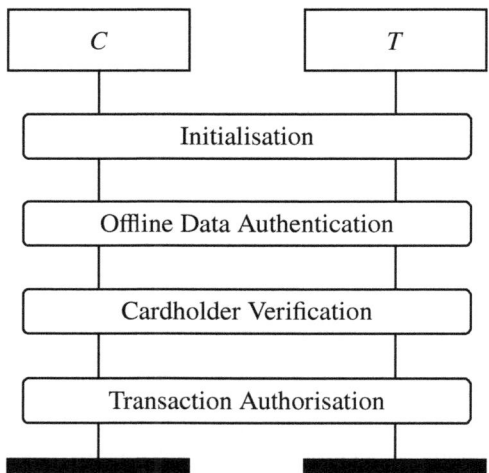

Fig. 2.8 The EMV protocol stages.

A transaction consists of at most four phases, as depicted in Fig. 2.8. Of these four phases, only two phases are mandatory: Initialisation, where the card receives the transaction details, and Transaction Authorisation, where the card generates the AC. The other two phases are optional: Offline Data Authentication (ODA), where the terminal validates the card using the public key of the payment system, and Cardholder Verification, where the terminal verifies the cardholder via, e.g., the PIN entered in the terminal's pad.

2.3 Motivating case study 2: Threats to the EMV protocol

In all phases, certain information is required, for both the card and the terminal. The card permanently stores several digital certificates, the signing secret key sk_c, and the shared secret mk with the bank (often called a master key). However, only the following is mandatory according to the standard [3]:

- Application Expiry Date
- Application Primary Account Number (PAN), i.e., the card number
- Card Risk Management Data Object Lists (CDOLs)

Other information the card may store will become clearer later. The terminal only stores the payment system's public key $\text{pk}(sk_s)$ used to verify the data on the card.

In the following four sections, we explain the four different (possible) phases of the EMV protocol in detail and highlight security and privacy concerns that immediately arise. This overview is intended to provide an intuition without many formal details, however, for the sake of completeness we recall cryptographic primitives the reader will encounter and provide the respective equations before proceeding.

> **Note on message syntax: asymmetric encryption and digital signatures.**
>
> *Asymmetric encryption* assumes three algorithms: public key pk, which produces a public key from a given secret key; encryption { }, which produces an encrypted message from a given message and a public key; and decryption adec, which recovers the message from a given encrypted message and the respective secret key. The equation involving the destructor adec is as follows.
>
> $$\text{adec}\Big(\{M\}_{\text{pk}(K)}, K\Big) =_\text{E} M$$
>
> where K plays the role of the secret key. Notice that syntax is typically overloaded in the literature for symmetric and asymmetric encryption. It should be clear from the context – e.g., the use of public keys indicates whether symmetric or asymmetric encryption is used. However, they should be treated as distinct primitives in tools.
>
> *Digital signatures* also assume three algorithms: public key pk, as above; signing sign, which produces a signature from a given message and a secret key; and checking check, which verifies the signature using the respective public key. Hence, two components are required for signature verification: the original message and the signature itself. The signature is verified if the message coincides with the result of applying the check function (that we treat as destructor) to the signature using the appropriate public key.
>
> $$\text{check}(\text{sign}(M, K), \text{pk}(K)) =_\text{E} M$$
>
> In what follows, we call the message-signature pair a *certificate*. Notice that we have used $=_\text{E}$ instead of $=_\text{D}$ to distinguish from the theory used in the previous case study for modelling symmetric encryption.

2.3.2 Initialisation

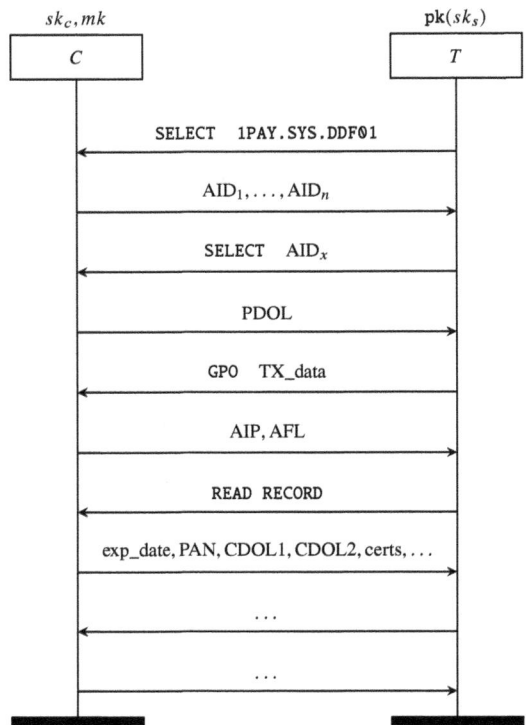

Fig. 2.9 Initialisation of the EMV protocol.

The initialisation phase is presented in Fig. 2.9. Firstly, the terminal asks which applications the card supports by issuing the SELECT command with a payload depending on the transaction type[4]. If the card is inserted into the terminal, i.e., the transaction is contact, then the 1PAY.SYS.DDF01 message is added to the command. If the transaction is contactless, 2PAY.SYS.DDF01 is added.

Here, Fig. 2.9 depicts the first option. The card responds with a list of payment application identifiers (AID), e.g., Visa Debit, Maestro, etc. Then, the terminal selects a particular application from the received list.

Having selected the AID, the card sends the Processing Options Data Object List (PDOL), specifying transaction details the card requires from the terminal. This

---

[4] Issuing a command with a payload is encoded as a pair (COMMAND, PAYLOAD), written informally in message sequence charts as COMMAND PAYLOAD.

2.3 Motivating case study 2: Threats to the EMV protocol

includes e.g., the amount, the date, and the currency. Next, in one message, the terminal sends the requested transaction data specified in PDOL and issues the Get Processing Options (GPO) command to request the Application Interchange Profile (AIP), the list of functions supported by the card including authentication methods, and the Application File Locator (AFL), the list of memory addresses where the card stores the data needed to complete the transaction. Upon receiving the AIP and the AFL from the card, the terminal issues the READ RECORD command and uses the addresses from the AFL to read the actual data from the card and immediately checks if the card is valid at the time of the purchase using the received expiry date. The rest data is used later, e.g., the card number PAN is used to route the *AC* through the network to the issuing bank at the last phase of the transaction. The CDOLs specify the information the card needs from the terminal to generate the application cryptogram *AC*, e.g., the country code, the terminal nonce, etc. Typically, certain certificates, i.e., digitally signed data, that the terminal should check using the appropriate public key is also among the data indicated by the AFL. Let us briefly explain this data.

The goal of the terminal is to ensure that the card is authentic, i.e., it was issued by a legitimate bank in collaboration with a payment system that the terminal supports. This is done by verifying the so-called *chain of certificates* called certs in Fig. 2.8. The chain of certificates is comprised of two ingredients.

- $(\text{pk}(sk_b), \text{sign}(\text{pk}(sk_b), sk_s))$, the certificate on the issuing bank's public key $\text{pk}(sk_b)$ issued with the private key of the payment system sk_s
- $(\text{pk}(sk_c), \text{sign}(\text{pk}(sk_c), sk_b))$, the certificate on the card's public key $\text{pk}(sk_c)$ issued with the private key of the issuing bank sk_b

After verifying the first certificate using $\text{pk}(sk_s)$, the terminal knows that the issuing bank's public key $\text{pk}(sk_b)$ provided by the card is authentic and can be used safely to verify any possible data on the card certified by the issuing bank including the card's public key in the second certificate. This design allows for setting up new terminals easily, as only the public key of the payment system needs to be stored.

Notice, however, that none of the data provided by the card is authenticated at this initial stage of the transaction, and we cannot yet reason about the security of payments. The privacy of the cardholder, on the other hand, is already violated since a lot of sensitive data is exposed to the environment in cleartext. Data elements that are unique only to a group of cards, such as the list of supported applications or the AIP, contribute to the card's *fingerprint*, thus enabling profiling. Data elements that are unique to a card, such as the PAN or the chain of certificates, can be used to link sessions made with the same card, thus violating both anonymity and unlinkability.

Notice that there are two types of attackers when it comes to EMV payments. A passive attacker can only observe communications and profile cardholders engaged in transactions, yet an active attacker, in addition, can power up a contactless card without the cardholder being aware by placing an antenna, e.g., at a doorway or by a seat on public transport. A powered-up card is ready to start an EMV session and to present its PAN – a strong form of identity. This enables an active attacker to silently track the movements of anyone who holds a payment card, even without a genuine EMV transaction being involved.

The ability to secretly activate a contactless card and communicate with it is called *skimming* in the literature. A skimming attack may be part of a relay attack or may serve as the basis of an attack on unlinkability, as described above. The skimming attack is possible within close distance to the card – Habraken et al. constructed an antenna in the form of a gate, up to 100cm in width, that can power the card and communicate with it [63]. For a passive counterpart of skimming, an eavesdropping attack that does not require powering up a contactless card, Engelhardt et al. achieved a distance of almost 20m [53].

Such total lack of privacy should not come up as a surprise, as the primary objective of EMV is the security of money in the cardholder's account, not privacy. However, even to achieve this primary goal, one should carefully select a secure configuration of the EMV protocol and avoid insecure ones. At the end of this chapter we will demonstrate an insecure configuration that is still permitted by the standard.

2.3.3 Offline Data Authentication

Offline Data Authentication (ODA) is an optional phase in the EMV protocol during which the terminal authenticates the data previously received from the card during the Initialisation phase. The ODA can be conducted in three different modes. We briefly describe each.

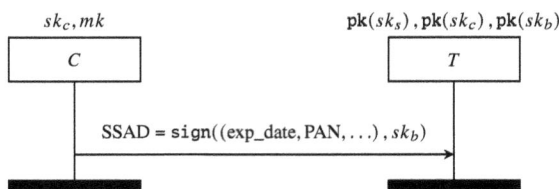

Fig. 2.10 The Static Data Authentication mode for Offline Data Authentication

Static Data Authentication (SDA) is schematically presented in Fig. 2.10. In this mode, the terminal uses the bank's public key to check the signature on the card's data retrieved by the terminal at the initialisation step (indicated by the AFL) issued with the bank's secret key sk_b. Security-wise, the authentication property called *aliveness* fails in this mode: the signature is the same for all sessions using SDA; thus, even after completing a session, it is not necessarily the case that the real card was present and communicating (alive), thereby card cloning (both the card's

2.3 Motivating case study 2: Threats to the EMV protocol

data, and the signature is exposed) can trick the SDA mode. Privacy-wise, since the signature is unique to a card, it can be used to link sessions with that card.

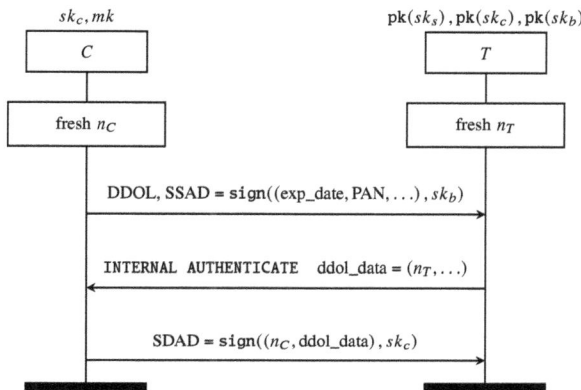

Fig. 2.11 The Dynamic Data Authentication mode for Offline Data Authentication.

Dynamic Data Authentication (DDA) mode is presented in Fig. 2.11. It mitigates the card cloning vulnerability of SDA and is essentially a terminal's check of a signature generated by the card over certain dynamic data. At the Initialisation phase, the card could provide the Dynamic Data Authentication Data Object List (DDOL) indicating the data elements that the terminal must send to the card if DDA is selected. The only mandatory element that the terminal should provide (even if the DDOL is not present) is a nonce n_T. To perform DDA the terminal firstly authenticates the AFL data, as in the SDA mode, secondly, it provides the DDOL data and issues the INTERNAL AUTHENTICATE command, and thirdly, it uses the card's public key to check the signature generated with the card's secret key sk_c over the provided DDOL data and the card's dynamic data which typically includes the card's nonce n_C. In EMV terminology, the data with the card's signature over it is called Signed Dynamic Authentication Data (SDAD). In contrast to SDA, the signature checked in DDA uses the session-specific data generated by both agents during the session, hence cannot be forged and used in card cloning. From a privacy perspective, the DDOL (or the lack of) identifies the card to a degree and contributes to the card's fingerprint.

Combined Data Authentication (CDA). This mode is similar to DDA, however, it does not require a specific message exchange, since CDA is a part of the Transaction Authorisation phase explained below. The card, in this case, generates the signature over the card's nonce and transaction-specific data, e.g., the cryptogram *AC*.

We conclude by summarising or main security and privacy observations for this phase of an EMV transaction. The SDA mode is vulnerable to card cloning while both DDA and CDA mode are not. Each ODA mode requires data unique to a card being revealed: in case the SDA mode is selected, the static signature that includes the mandatory PAN is unique, and in case the DDA or CDA mode is selected, the card's unique public key (transferred at the initialisation) is required to check the card-generated signature. Given that the communication is in cleartext, even an eavesdropper can recognise if the same card is used across different sessions and, thereby, track the cardholder.

2.3.4 Cardholder Verification

As the name suggests, this phase shall prove that the person using the card is indeed a legitimate cardholder. Like the previous phase, this phase is optional. Cards that support cardholder verification provide Cardholder Verification Methods (CVM) list CVM_list to the terminal in the Initialisation step in a standard way – the card indicates the address of the CVM_list in the AFL, and then the terminal reads it using the READ RECORD command. One major security issue here is that it is not required that the CVM_list is part of the static data to be authenticated. Hence, if not authenticated, it can be altered by a man-in-the-middle attack with the goal to bypass cardholder verification.[5] Privacy-wise CVM_list differs from card to card from a privacy perspective, thus contributing to its fingerprint, which becomes more distinct at this stage than at the Initialization. Note that in some situations, the fingerprint may serve as a unique identifier for the card, e.g., a card from abroad is likely to be configured differently than most local cards and it is easy to link payments made with it based on the EMV flavour it employs.

Verification methods include handwritten signature, Personal Identification Number (PIN), and verification via the consumer's device (e.g., through biometric data entered via a mobile phone), which is out of the scope of the standard. Interestingly, the Payment System determines who is held liable for disputed transactions based on the CVM used. If the cardholder was verified using their signature, the merchant is liable, if a PIN has been used, the cardholder is liable, and, in the case of, e.g., Mastercard, if CVM is skipped entirely, the acquiring bank bears the liability [7]. Now we consider several possible PIN-based CVMs.

Offline Cleartext PIN. In this mode, the PIN, entered into the terminal, is sent to the card in clear text, together with the VERIFY command as indicated in Fig. 2.12. If the PIN is correct, the card responds with a constant message indicating success. Otherwise, the card responds with 63Cx message, where x is the number of tries left. The first security threat here is very obvious. The PIN is exposed to eavesdroppers, making it straightforward to use the stolen/cloned card or relay messages from someone's pocket to a terminal to make a payment. We stress the passive nature of the eavesdroppers here, as entering the PIN assumes a user action and the awareness of the cardholder being in the middle of a purchase. In contrast, the PIN is not among the data that an attacker secretly activating the card can silently gather. The second

---

[5] The card brand mixup attack [24], however, demonstrates that even if this list is authenticated, it can be downgraded to offline cleartext PIN-only list.

2.3 Motivating case study 2: Threats to the EMV protocol

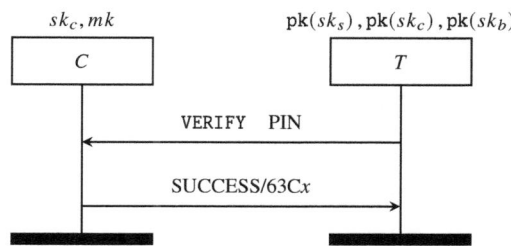

Fig. 2.12 CVM: Offline cleartext PIN.

security threat is slightly more subtle. The card's response is not authenticated, making it possible for a man-in-the-middle attacker to block this response from the card and to lie about the PIN verification outcome. Privacy-wise, while the PIN is not necessarily unique to a card, the likelihood of transactions in which the same PIN is used are made with the same card is high, hence the PIN could be used to link sessions and track the cardholder. Moreover, if the entered PIN is wrong, the card reveals the number x of tries left and becomes temporarily distinguishable among others – with a good probability an EMV protocol session in which x, and a session in which $x - 1$ is revealed, are made with the same card, given these sessions are at approximately the same location.

Offline Encrypted PIN. The above PIN leak vulnerability can be solved by encryption, as shown in Fig. 2.13. In this method, the terminal firstly issues the GET CHALLENGE command to the card, and then uses the public key of the card $\text{pk}(sk_C)$ to encrypt the received card's nonce n_C and the PIN. The verification result is then received unauthenticated again though. This CVM mode indeed hides the PIN from the eavesdropper and also makes the encryption of the PIN local to the current session, thus avoiding the replay attack and making it impossible to use the encrypted PIN to link sessions. However, the response from the card can still be blocked and replaced with SUCCESS even if the entered PIN was wrong.

Online encrypted PIN. Finally, if the terminal can perform online transactions instead of the card, the (encrypted) PIN is transmitted to the issuing bank for verification together with the cryptogram *AC* and other required data. In this case, the CVM is part of the Transaction Authorisation phase. Since, in this chapter, we focus on the communication between the card and the terminal (which is generally less protected than the terminal-bank communication), we omit the detailed description of this method.

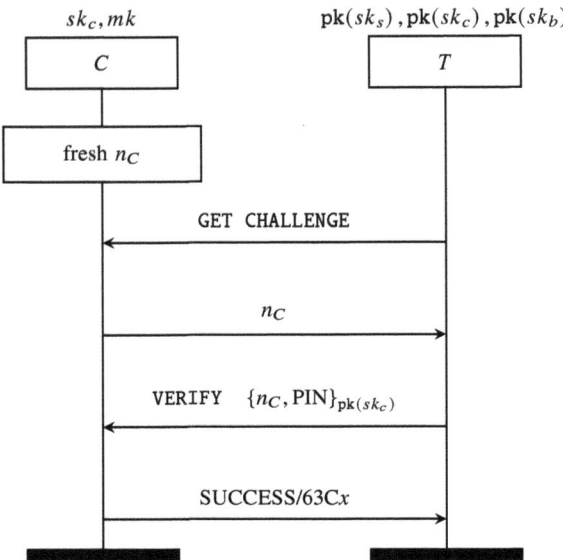

Fig. 2.13 CVM: Offline encrypted PIN mode. The digit x is the number of tries left.

2.3.5 Transaction Authorisation

Transaction Authorisation (TA) is the final and mandatory phase at which the terminal asks the card to generate the Application Cryptogram AC. Depending on the transaction type, online or offline, the cryptogram is sent to the issuing bank during or after the TA phase. The bank then verifies the cryptogram and replies to the terminal appropriately. There are three types of cryptograms that the terminal may ask the card to generate.

- Transaction Cryptogram TC, a success cryptogram
- Authorisation Request Cryptogram $ARQC$, a request for online confirmation to the issuing bank
- Application Authentication Cryptogram AAC, a cryptogram used to log the failed attempt

A cryptogram is typically a MAC, generated over the data coming both from the card and the terminal. The data provided by the terminal is specified by CDOL objects, which the terminal has received in the Initialisation phase. The key for this MAC is derived from the Application Transaction Counter (ATC), a counter on the card that increments each time the GET PROCESSING OPTIONS command

2.3 Motivating case study 2: Threats to the EMV protocol

is issued[6], and the master key mk, shared between the card and the issuing bank. In short, the application cryptogram follows the following pattern, where f is some mechanism of deriving a key.

$$AC = \mathtt{mac}((\ldots, \text{ATC}, \text{ddol\_data}, \ldots), f(\text{ATC}, mk))$$

Notice, that it is only the issuing bank that can verify the cryptogram – the terminal has no access to mk and simply forwards the data that goes into the cryptogram and the cryptogram itself to the issuer. Also, notice that the use of monotonically increasing counter ATC prevents the re-use of old cryptograms. The minimum set of data elements to be included in the cryptogram recommended by the EMVCo [2] contains, for instance, ATC, AIP, terminal country code, transaction date, amount, etc. As this data is not encrypted when sent from the card to the terminal and may contain identifying information about the card (e.g., the counter ATC), linkability issues may arise.

Before explaining both offline and online modes, we clarify how the CDA mode for offline data authentication works, as it is part of the Transaction Authorisation step. Recall that the Dynamic Data Authentication (DDA) mode requires the card to sign the dynamic data coming both from the card and the terminal, i.e., nonces n_C and n_T. The CDA is very similar, but together with n_C and n_T the card also signs the cryptogram, the Cryptogram Information Data (CID) which specifies the type of the cryptogram, and the hash of the transaction details. This signature is only generated when the requested cryptogram is TC or $ARQC$ and it is never generated if the requested cryptogram is AAC. In what follows, we assume that CDA is either not selected for offline authentication of the card's data or ODA is skipped. Now we describe how offline and online transactions are finalised.

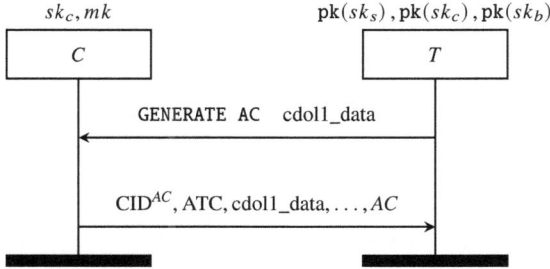

Fig. 2.14 Offline transaction authorisation.

---

[6] Notice, that an (active) attacker can use a device capable of communicating with the card to issue enough GET PROCESSING OPTIONS commands to max out the two-byte ATC and, hence to disable the card.

Offline authorisation. We present the offline TA in Fig. 2.14. The terminal asks the card to generate the cryptogram *AC* by issuing the GENERATE AC command together with the data indicated in CDOL1, the payload indicating the type of the requested cryptogram and a parameter indicating whether the CDA is requested. The card responds with the CID^{AC} indicating the type of the cryptogram, the counter ATC, the cryptogram $AC = \text{mac}\big(\big(\text{CID}^{AC}, \text{ATC}, \text{cdol1\_data}, \ldots\big), f(\text{ATC}, mk)\big)$, and, optionally, proprietary Issuer Application Data (IAD) to be sent to the issuing bank. The card can reply with either the cryptogram of the requested type, *ARQC* or *AAC*. Though the terminal does not verify the received cryptogram *AC*, and sends it to the issuing bank later, an offline transaction at this stage is complete.

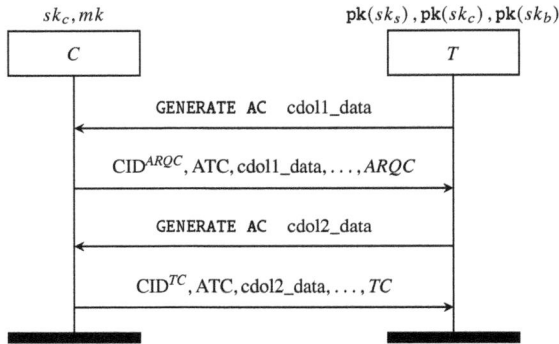

Fig. 2.15 Online transaction authorisation.

Online authorisation. Online TA proceeds as in Fig. 2.15. In an online transaction, two cryptograms are generated. Firstly the terminal requests the card to generate an *ARQC* cryptogram by issuing GENERATE AC and providing CDOL1 data, the cryptogram type and the CDA flag similarly to the offline mode. Then the card replies, and the terminal immediately requests the issuing bank to authorise the current transaction by forwarding the

$$ARQC = \text{mac}\big(\big(\text{CID}^{ARQC}, \text{ATC}, \text{cdol1\_data}, \ldots\big), f(\text{ATC}, mk)\big)$$

to the bank, together with the transaction details and the (encrypted) PIN in case online PIN CVM was selected. If the confirmation from the bank and further data[7] has been received, the terminal issues the second GENERATE AC command to the card asking the card to generate the *TC* and providing CDOL2 data which includes the

---

[7] The data received from the bank may also contain instructions for the card to update certain data inside it. Despite being part of the online TA, in EMV, this script processing step is often regarded as a separate optional phase of the transaction.

2.3 Motivating case study 2: Threats to the EMV protocol   41

bank's response. When the $TC = \mathtt{mac}\bigl(\bigl(\mathrm{CID}^{TC}, \mathrm{ATC}, \mathrm{cdol1\_data}, \ldots\bigr), f(\mathrm{ATC}, mk)\bigr)$ is received by the terminal, the transaction is completed. In case the banks rejects the request, the terminal still asks the card to generate the rejection cryptogram *AAC*.

Contactless transactions. In the above, we have described contact transactions. The main difference with the contactless case is that the application cryptogram is generated right away when the transaction data is received and is sent to the terminal in response to the GET PROCESSING OPTIONS command (see the description of the Initialisation phase above). Moreover, neither user input nor reply from the bank can be received by the card, as the reader should keep the card within the reader's field throughout the whole session. In this case, the same security and privacy implications apply, keeping in mind that a passive eavesdropper can gather exposed data from a distance and an active attacker can activate and communicate with the card without being noticed.

As the reader may observe by now, the EMV standard is flexible. A transaction contains a variety of optional messages that the card and the terminal may or may not include in the communication. Moreover, Offline Data Authentication and Cardholder Verification can be skipped if a payment system decides to implement the "reduced" version of the standard. Formulations such as "the recommended minimum set of data elements to be included in . . .", " it is strongly recommended . . .", etc., common for EMV books, suggest that deviating from recommendations is allowed since there are also explicit mandatory requirements. Insecure configurations to this date are part of the EMV standard, as has been demonstrated in [25] where the authors presented a full message sequence chart of the EMV protocol for the first time, and analysed 24 EMV configurations for contactless transactions among which only three have been found secure. One such configuration has been independently discovered by the authors of this book by studying the EMV standard with the naked eye and is presented below.

2.3.6 Examples of insecure EMV configurations

In the earlier sections, we explained the main phases of the EMV protocol. In this section, we combine different observations, yielding EMV configurations that comply with the EMV standard but are still vulnerable to a PIN bypass attack.

We start with a long-known vulnerability disclosed by Murdoch et al. [94] in Fig. 2.16, rooted in the following observations.

- The CVM_list is not required to be among the data authenticated by the terminal during the ODA phase, hence it can be altered by the man-in-the-middle attacker
- The card's response in the offline PIN verification is not authenticated either

To execute an attack, first, an attacker E, if necessary, downgrades the options of cardholder verification methods not to contain online verification by modifying the CVM_list. Second, upon receiving (any) PIN, it simply blocks it from reaching the card and replies with the message that represents success. From the exchanged messages, the terminal will think the PIN was correct while the card will determine the signature option was chosen since it did not receive any VERIFY command. The remaining messages are simply forwarded.

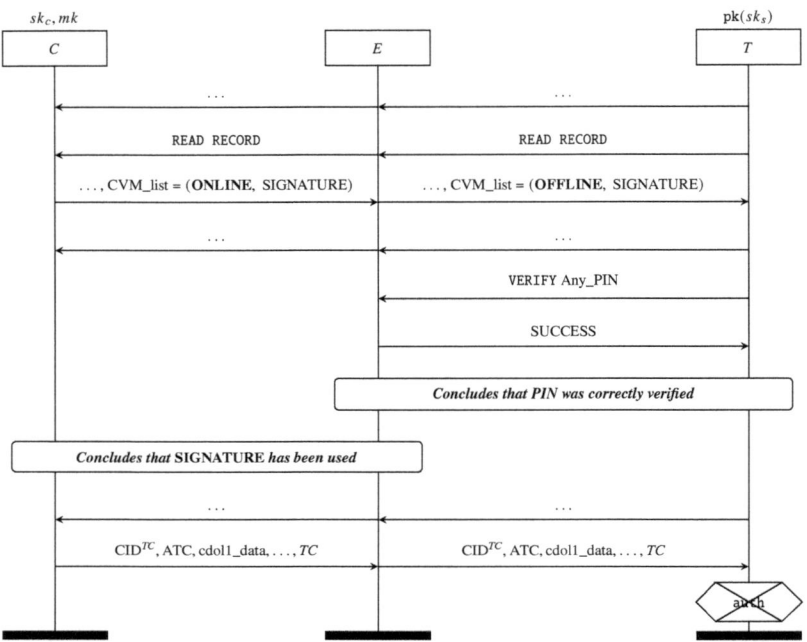

Fig. 2.16 Long-known vulnerability of EMV protocol [94]. The attacker blocks the entered PIN from reaching the card and sends back the SUCCESS reply. The card receives no VERIFY command, concludes that the terminal does not support PIN CVM, and that the cardholder was verified by the signature.

The attack that we present next is of the same nature and the authors of this book are not aware if it has been reported before. This attack is theoretical and can be executed only for cards configured in a very particular way, we can neither guarantee the existence of cards which carry this configuration (hence vulnerable to the attack) nor the opposite. The point we are stressing in this section is that insecure configurations are still part of the EMV standard, while they should not be allowed in the first place.

Recall that the CVM_list is not required to be included in SSAD, the data to be authenticated by the terminal, hence, if not, it can be altered by the man-in-the-middle attacker. Let us first make an additional observation regarding the cardholder verification phase.

2.3 Motivating case study 2: Threats to the EMV protocol

- The PIN is sent to the card in case of offline PIN verification, while it is sent to the bank in case of online PIN verification

Hence, the strategy of an attacker is to introduce the disagreement between the online and offline modes as presented in Fig. 2.17.

Fig. 2.17 Overview of the strategy of an attacker.

To develop this strategy into a full attack, we assume particularly configured card and terminal. The CVM_list on the card should not be included in the SSAD (the static data that the terminal authenticates) and should comprise a unique Cardholder Verification Method, online PIN. The terminal should be offline and support offline PIN verification. Given these assumptions, the full attack from Fig. 2.18 can proceed as follows.

At the Initialisation phase, the man-in-the-middle attacker replaces the online PIN entry in the CVM_list with any version of the offline PIN verification, e.g., offline encrypted PIN. The terminal then executes the offline CVM while the man-in-the-middle blocks any messages regarding the PIN verification from reaching the card and ultimately responds with the SUCCESS message. Then the terminal completes the offline transaction asking the card to generate the *TC* allowing an attacker to successfully use the card without knowing the PIN. Recall that the cardholder would be liable for this fraudulent transaction as the PIN-based CVM has been used.

Finally, we make an observation that the card never authenticates the terminal and its choice of the CVM. Hence, in case CVM_list = (ONLINE, OFFLINE) is authenticated, i.e., is included in the SSAD, the described attack is still possible.

Further examples of attacks on EMV leading to fraudulent payments include not only PIN bypassing [94, 25, 24], but also downgrade attacks [94] and pre-play attacks [33, 32]. The majority of such attacks, as we have just observed, are rooted in the failure of the selected EMV configuration to deliver authentication guarantees, i.e., messages between the agents can be altered on the way, leading to different views of the same transaction from the perspectives of the card and the terminal.

In contrast, relay attacks [49, 35], where messages remain unaltered but are simply relayed (e.g., between a card in a wallet and a terminal), stand out and require the introduction of distance-bounding techniques [43, 85], ensuring that the agents are close in order to execute the transaction. We will study such protocols in Chapter 6. Finally, relaying can be combined with a man-in-the-middle attack to execute a fraudulent payment, as demonstrated by, e.g., Radu et al. in [99], where they describe how to bypass the Apple Pay lock screen when using a Visa card.

This section completes the description of the current EMV standard and highlights its security and privacy issues. Ill-configured versions of the protocol still exist in

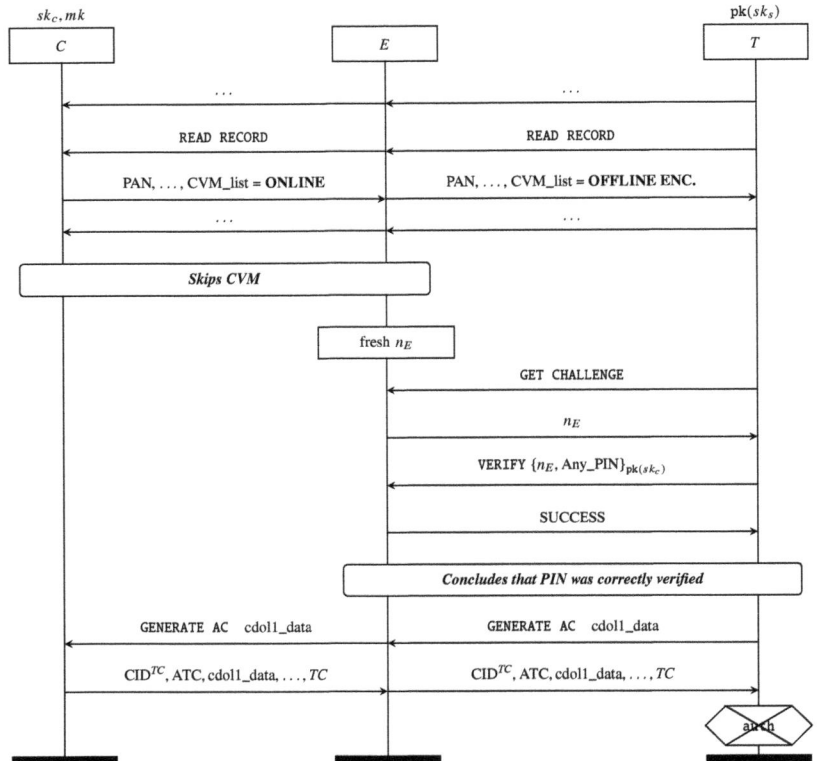

Fig. 2.18 Attack bypassing the PIN in an offline transaction.

the standard without any indication that they should never be implemented. Sensitive data allowing either identification or profiling of cards is exposed to eavesdroppers. Relatively recently, EMVCo proposed enhancing the privacy of transactions and mitigating eavesdroppers by encrypting the communication between the card and the terminal [5], which has since been adopted in EMV Kernel 8 [52]. We will analyse the privacy of this proposal towards the end of this book.

The purpose of this section is to motivate the reader with an example of a protocol that the majority of citizens, internationally, use on a daily basis in a critical fashion for which the security properties we investigate are evident. Indeed, one of the authors who discovered a recent attack [25] was a teaching assistant for the course on which this book is based, and the methods employed are based on an update of the same essential knowledge of symbolic verification of protocols. Why would experienced engineers make these design mistakes in such critical infrastructure? A possible answer is that security professionals really require methodologies for designing and verifying security protocols, such as those communicated in this book.

Part II
Modelling Protocols and Threats

Chapter 3
Syntax and operational semantics for security protocols

Abstract This chapter introduces an operational semantics for a core of the applied π-calculus, sufficient to capture a range of protocols and threat models. The basics of syntax, binders and substitutions are explained. The syntax of threads modelling the behaviour of an individual role in a protocol is explained in relation to message sequence charts modelling protocols globally. An operational semantics is provided for threads, defining how honest agents behave in an adversarial environment. Secrecy is defined and a modal logic for expressing attacks is introduced. The operational semantics are then extended to networks where many threads may run concurrently. Exercises help the reader gain a feel for the rules of the operational semantics.

3.1 Introduction

With this chapter, we start the formal treatment of security protocols. For that, we first need a syntax to specify security protocols – basically a formal syntactic representation of the security protocol – for which we use a process calculus. The operational semantics then gives meaning to these representations, which makes precise how protocols interact with their, possibly compromised, environment. This allows us, in subsequent sections, to define precise threat models and security goals that should be achieved in the face of those threats.

The operational semantics of the process calculus that this section is based on – the applied π-calculus – is undeniably non-trivial to understand. There are many process calculi with simpler semantics. However, all the ingredients presented here are necessary to understand security protocols: including inputs and outputs involving cryptographic messages, explicit secret information such as keys and nonces, and conditional statements. All of these ingredients come together to express how protocols interact with a network on which there may be an attacker that eavesdrops on outputs and induces inputs using their knowledge. A good deal of machinery is therefore required to ensure that the operational semantics encompasses the capabilities of such attackers. The chapter introduces those essential ingredients for manipulating syntax to define a state-of-the-art operational semantics.

How to read this chapter: This chapter sits alone, since it is possible to proceed with defining protocols, knowing only the syntax of processes and having an intuition for what those processes mean in relation to security properties. However, better understanding an operational semantics of protocols, helps form a much stronger intuition allowing us to understand what is really happening, for example, when a ciphertext is output during a protocol execution. This chapter may either be read linearly with the flow of the book or used as a reference when aiming to understand processes in future chapter more precisely. By having a feel for the operational semantics of the applied π-calculus the reader may better appreciate how hard the semantics works for us when we use the process syntax in tools like ProVerif.

3.2 Symbolic reasoning on syntax in computer science

Syntax is central to the philosophy of computer science. This is an aspect in which sense that the traditions of computer science differ from those of mathematics. In mathematics, typically we assume that the objects of study can be defined in terms of sets (or more abstract frameworks such as categories). There is also an established mathematical notation that has evolved over hundreds of years, for example, the work of Stefan Banach on functional analysis in 1932 uses much the same notation as found in modern day textbooks on the topic.

The notational uniformity found in mathematics is not prevalent in computer science. In computer science, we are used to the fact that there are multiple programming languages and data formats, each with their own advantages. This is due to one of the most powerful phenomena at our disposal – *abstraction*. When programming, choosing the right level of abstraction can help us more quickly or accurately express a problem. It is inappropriate to encode all problems in assembly, or C, or even Python. For some applications, functional programming languages are appropriate and, for others, we employ dedicated domain specific scripting languages. In the security domain, if we focus on modelling and analysing social-technical attacks we will use a different set of languages and tools compared to those that we would use if we aim to analyse the information flow and leakage in an Android application. In order to work efficiently in each domain and problem, without becoming distracted by routine implementation details, it is important to choose an appropriate level of abstraction for modelling the problem and, for that, the choice of syntax can help.

We should distinguish between concrete and abstract syntax. Concrete syntax is typically the string of characters that we feed to a compiler or another tool, which is designed for humans to type and for computers to parse. Internally, a lexer and parser turn this stream of characters into abstract syntax, which is a tree indicating how all the sub-terms of the calculus are defined. Abstract syntax gets rid of comments, line breaks and other sugar such as abbreviations, and makes brackets implicit from the tree structure of the syntax.

3.2 Symbolic reasoning on syntax in computer science

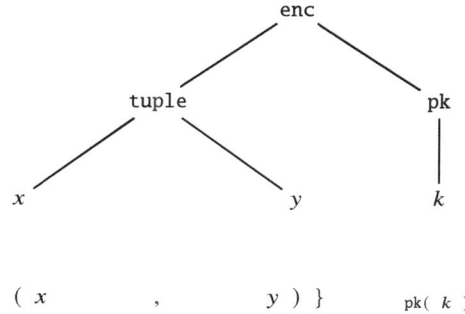

Fig. 3.1 An example of an abstract syntax tree parsing a string according to our grammar for messages.

When defining a model, it is efficient to define the abstract syntax first, and then only worry about the concrete syntax later. When working on modelling programs and systems, syntax is not just an input format for a tool, syntax also plays a role in modelling the state of a system as it is executed, i.e., we are using syntax as a way to represent a rich data structure. Thus, when confronted with a piece of syntax, we should not be tempted by the first thoughts of traditional mathematician which might be: what does it mean in terms of its denotation as a set. Instead, our first question should be what does the syntax mean in terms of it relationships with other syntactic terms, such as how do we move from one state to another during execution, which will prepare us for the thinking behind the field of structural operational semantics.

3.2.1 A syntax for messages and threads

Abstract syntax can be defined using a context-free grammar [67]. The standard notation we use in computer science for defining a context-free grammar is Backus-Naur form (BNF), named after two pioneers in the field of programming languages who introduced the notation to describe one of the earliest programming languages with a precise specification [19]. Both Backus and Naur have been recipients of the Turing award — one of the highest accolades in computer science.

We begin with a message language containing the common functions in security protocols. The **message terms** are defined using the grammar below. The example theory covers tuples and asymmetric encryption.

$$\begin{array}{lll}
\mathsf{M,N} ::= & x & \text{variable} \\
& \mid \mathsf{pk}(\mathsf{M}) & \text{public key} \\
& \mid \{\mathsf{M}\}_\mathsf{N} & \text{(asymmetric) encryption} \\
& \mid \mathsf{dec}(\mathsf{M},\mathsf{N}) & \text{(asymmetric) decryption} \\
& \mid (\mathsf{M},\mathsf{N}) & \text{tuple} \\
& \mid \mathsf{fst}(\mathsf{M}) & \text{first projection} \\
& \mid \mathsf{snd}(\mathsf{M}) & \text{second projection}
\end{array}$$

(SYNTAX OF MESSAGES)

In BNF form, each line of the grammar corresponds to a rule in a context-free grammar that can be applied to generate an abstract syntax tree that parses strings permitted by the language. For example, the abstract syntax tree presented in Fig. 3.1 accepts the string $\{(x,y)\}_{\mathsf{pk}(k)}$. We use that string to represent the encryption of a pair of variables x and y using the key $\mathsf{pk}(k)$ which is the public key corresponding to the private key k. We often abbreviate such encrypted tuples as $\{x,y\}_{\mathsf{pk}(k)}$, and abbreviate nested tuples, e.g., $(m_1,(m_2,m_3))$ as (m_1,m_2,m_3). This example syntax is a starting point that will be extended throughout the book as we introduce new concepts that require new syntax, such as signatures, hash functions, or XOR operators.

Remark 3.1 The notation $\{\mathsf{M}\}_\mathsf{N}$ for encryption is commonly used in the literature on security protocols. Notice that, in Section 2.2, symmetric encryption used the same notation. This overloading of notation is common, but it should be explicit from the context whether a public key or symmetric key is used to encrypt the message. To simplify matters, throughout this chapter and the next only public key cryptography is used.

We define a syntax for **processes**. Initially we specify only the threads that each agent participating in a protocol executes, and we will extend the syntax to cover more processes as the book progresses. By a thread we mean the linear order in which a single agent performs actions such as inputs and outputs, indicating which action comes before another action.

The syntax also features control constructs such as `if` statements as well as a fresh name binder for declaring nonces and fresh keys which are generated at runtime, such that they are different from any name previously generated (modelling a cryptographically random number or group element that is fresh with sufficiently high probability). Such fresh names are local to the scope of the binder and may be referred to as *names*, or by their role in the protocol, e.g., nonces or key material.

$$\begin{array}{lll}
\mathsf{T} ::= & 0 & \text{end} \\
& \mid \mathsf{secret}(\mathsf{M}) & \text{secret} \\
& \mid \mathsf{out}(\mathsf{M},\mathsf{N});\mathsf{T} & \text{send} \\
& \mid \mathsf{in}(\mathsf{M},y);\mathsf{T} & \text{receive} \\
& \mid \text{if } \mathsf{M} = \mathsf{N} \text{ then } \mathsf{T} & \text{match} \\
& \mid \text{if } \mathsf{M} = \mathsf{N} \text{ then } \mathsf{T}_1 \text{ else } \mathsf{T}_0 & \text{conditional} \\
& \mid \mathsf{new}\, x;\mathsf{T} & \text{fresh} \\
& \mid \text{let } x = \mathsf{M} \text{ in } \mathsf{T} & \text{let}
\end{array}$$

(SYNTAX OF THREADS)

3.2 Symbolic reasoning on syntax in computer science   51

The action out(M, N);T represents an action of the agent in the session represented by the thread, where the agent sends a message N on the network, whatever that might be, along the channel M. At the moment we do not need to know what a channel is, and it will be used later in this book creatively, but we can see it for now as simply an identifier for a port from which the message is sent. After having sent the message, the thread proceeds with the rest of the thread represented by T in the grammar. The action in(M, y);T similarly represents an input action on some channel labelled M, before proceeding with the thread. The difference is that, with an input the message received is represented by a variable, say y, which is a placeholder for whichever message is received from the network when the input occurs, and will be used in T. If-then-else statements are also supported, where if M ≠ N, then the thread in the else-branch is taken. Thread terms of the form if M = N then T represent a conditional where if M and N are equal then the thread will proceed, otherwise the thread aborts, behaving like 0, which can be seen as an abbreviation of if-then-else. The let _ in _ statement is included simply to make processes more readable by abbreviating a larger message with a variable.

Threads can either end with 0, indicating we have reached the end of the thread, hence there are no further actions to perform, or it ends with a *secret* event. The *secret* event is an assertion that may only appear at the end of thread, which we will used to specify secrecy properties by asserting that a message that should be secret is never revealed to an attacker. A syntactic convention we adopt is that we omit the final end thread, so, for example, the thread out(c, n);0 is written out(c, n).

We have chosen the notation for operators such that it is both concise and intuitive to read. This choice of syntax aligns closely with the syntax used in the ProVerif tool (and related tools such as SAPIC+ [40] and DeepSec [41]), which is just an ASCII syntax for the applied π-calculus. In Fig. 3.2, we compare the ProVerif syntax to the traditional applied π-calculus syntax, and also Scyther mentioned previously.

ProVerif	applied π-calculus	Scyther
new x; T	νx.T	**fresh** x: Nonce; T
in(M, y); T	M(y).T	**recv_l**(i, r, N); T
out(M, N); T	$\overline{M}\langle N\rangle$.T	**send_l**(i, r, N); T
if M = N **then** T	[M = N]T	**match**(M, N); T

Fig. 3.2 Comparing the ProVerif, applied π-calculus and Scyther syntax.

The applied π-calculus syntax is often used in publications, since it is the most concise established syntax for processes; allowing more complex examples to be expressed, but can be initially confusing for a new reader.

You can see that the notation does not diverge much. Only Scyther [47] uses a genuinely different style, that forces more information to be included than in the applied π-calculus, as we outline briefly here. Firstly, in Scyther, the events are uniquely labelled with distinct numbers, which is for internal purposes. Secondly, the sender and receiver, say i and r, are always named by Scyther, whereas the other notation makes use of channels, e.g. M. In these notes, when the sender or receiver should be made explicit, we include the routing information along with the message sent, say N, sent as a tuple say (i, r, N). Thirdly, the type Nonce is indicated explicitly when declaring a fresh name. Fourthly, instead of representing an input by a variable, Scyther performs pattern matching on two message terms to determine the inputs received and check that certain data appears in an input, which combines several operations in other models.

Despite these differences, Scyther does not differ significantly in terms of modelling security protocols compared to the applied π-calculus and ProVerif. Hence this book can be adapted to several process languages. The tool Tamarin [87] expresses protocols in a different style, that exposes more of details of the operational semantics. Tamarin is the subject of another book in this series [22], hence is not covered here.

3.2.2 Preliminaries on substitutions

We use substitutions to manipulate syntax by replacing variables with larger messages. For example, if $f(x) = x + 1$ then to calculate $f(2)$ we first take $x + 1$ and substitute x with 2 to obtain $2 + 1$. Similarly substitutions are used in processes, when realising inputs, by replacing the input variable with the message that is received during execution. We will see, as this chapter progresses, that substitutions are really central to modelling attacker knowledge. Substitutions will be used to map between the view of an attacker intercepting messages and the messages that are actually being exchanged inside threads. For this reason, it is worth investing time in knowing the basics of substitutions.

This section contains technical definitions for what it means to be a substitution. An important idea is that substitutions should be treated with care in the presence of local "bound" variables which are declared by a construct such as new x; T. Bound variables are treated differently from "free" variables which may appear elsewhere in terms. An important related concept is that some variables are "fresh" for terms, in the sense that it is safe to use the variable without conflicting with a previous use of the same variable for another purpose elsewhere in a term. These notions are defined precisely in this section.

3.2.2.1 Free variables

In syntax, some of the terms act as binders for variables. In particular, the fresh name binder and the variable representing an input bind occurrences of the variable that appear, which essentially make the variable local to the term, hiding it from the outside world. For example, in the thread "new k; in(c, x); out$\left(c, \{(x, y)\}_{\text{pk}(k)}\right)$" the prefix "new k;" binds the occurrence of k in the output. Also the variable x represent an input that is not yet known, which binds the occurrences of x. In contrast, the variables c and y are not bound by inputs or by fresh name binders in the above

3.2 Symbolic reasoning on syntax in computer science

thread. The binding structure for this example is visualised in Fig. 3.3, by annotating the abstract syntax tree with arrow from binders to the variables bound by the binder.

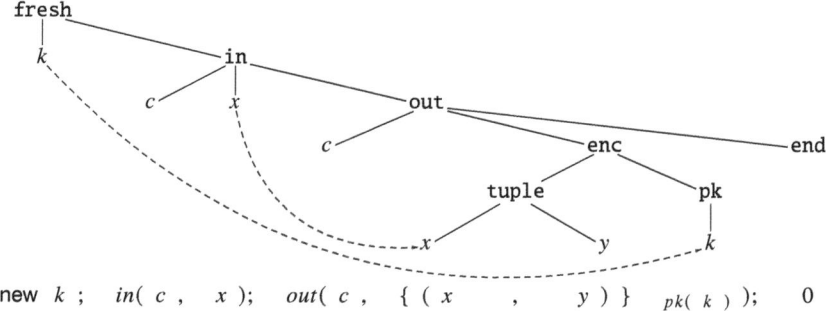

Fig. 3.3 Graphical representation of binders and bound variables.

The variables appearing in terms that are not bound are called free variables. Since there are no binders in messages themselves, the set of free variables in messages are defined by the following functions from messages to sets of variables.

$$\text{fv}(x) = \{x\} \qquad \text{fv}(\text{pk}(M)) = \text{fv}(M) \qquad \text{(Free variables of messages)}$$

$$\text{fv}(\{M\}_N) = \text{fv}(M) \cup \text{fv}(N) \qquad \text{fv}(\text{dec}(M, N)) = \text{fv}(M) \cup \text{fv}(N)$$

$$\text{fv}((M, N)) = \text{fv}(M) \cup \text{fv}(N) \qquad \text{fv}(\textbf{fst}(M)) = \text{fv}(M) \qquad \text{fv}(\textbf{snd}(M)) = \text{fv}(M)$$

The style of presenting functions over syntax above shows a case to apply for every syntactic construct. For example we have:

$$\begin{aligned}
\text{fv}\Big(\{(x, y)\}_{\text{pk}(k)}\Big) &= \text{fv}((x, y)) \cup \text{fv}(\text{pk}(k)) && \text{since } \text{fv}(\{M\}_N) = \text{fv}(M) \cup \text{fv}(N) \\
&= \text{fv}(x) \cup \text{fv}(y) \cup \text{fv}(\text{pk}(k)) && \text{since } \text{fv}((M, N)) = \text{fv}(M) \cup \text{fv}(N) \\
&= \text{fv}(x) \cup \text{fv}(y) \cup \text{fv}(k) && \text{since } \text{fv}(\text{pk}(M)) = \text{fv}(M) \\
&= \{x, y, k\} && \text{since } \text{fv}(x) = \{x\}
\end{aligned}$$

For a thread, the set of free variables is defined by the following function from threads to sets of variables. Notice that we here have two binders (new x; T and in(M, x);T) which remove the bounded variable x from the set of free variables in a thread.

$$\mathrm{fv}(0) = \emptyset \qquad \mathrm{fv}(\mathrm{secret}(M)) = \mathrm{fv}(M) \qquad \text{(FREE VARIABLES OF THREADS)}$$

$$\mathrm{fv}(\mathrm{out}(M,N);T) = \mathrm{fv}(M) \cup \mathrm{fv}(N) \cup \mathrm{fv}(T) \qquad \mathrm{fv}(\mathrm{in}(M,x);T) = \mathrm{fv}(M) \cup (\mathrm{fv}(T) \setminus \{x\})$$

$$\mathrm{fv}(\mathrm{if}\ M = N\ \mathrm{then}\ T) = \mathrm{fv}(M) \cup \mathrm{fv}(N) \cup \mathrm{fv}(T) \qquad \mathrm{fv}(\mathrm{new}\ x; T) = \mathrm{fv}(T) \setminus \{x\}$$

$$\mathrm{fv}(\mathrm{let}\ x = M\ \mathrm{in}\ T) = \mathrm{fv}(M) \cup (\mathrm{fv}(T) \setminus \{x\})$$

The predicate "$x \# T$", pronounced x is fresh for T, is used to indicate that the variable x does not appear free in the term T (whether T is a message, thread or some other term we will define). That is, we define freshness as follows.

Definition 3.1 $x \# T$ holds whenever $x \notin \mathrm{fv}(T)$.

This predicate extends point-wise to lists of entities. So, for example, $x, y \# T_1, T_2$ means: $x \# T_1, x \# T_2, y \# T_1$, and $y \# T_2$.

3.2.2.2 Substitutions

A substitution is a function that maps finitely many variables $x_1, x_2, \ldots, x_n$ to messages $M_1, M_2, \ldots M_n$ and maps all other variables to themselves. A single substitution is written $\{x \mapsto M\}$ and substitutions that change multiple variables can be defined by listing all their components. Thus $\theta = \{x_1 \mapsto M_1, x_2 \mapsto M_2, \ldots x_n \mapsto M_n\}$ defines the function from variables to messages as follows.

$$z\theta = \begin{cases} M_i & \text{if } z = x_i \text{ and } 1 \le i \le n \\ z & \text{otherwise} \end{cases}$$

Substitutions are defined on message terms by replacing occurrences of variables with messages. That is, substitution θ defines a function from messages to messages such that for any message M we define $M\theta$ as follows.

$$\mathrm{pk}(M)\theta = \mathrm{pk}(M\theta) \qquad \{M\}_N\,\theta = \{M\theta\}_{N\theta} \qquad \text{(SUBSTITUTIONS FOR MESSAGES)}$$

$$\mathrm{dec}(M,N)\theta = \mathrm{dec}(M\theta, N\theta) \qquad (M,N)\theta = (M\theta, N\theta)$$

$$\mathrm{fst}(M)\theta = \mathrm{fst}(M\theta) \qquad \mathrm{snd}(M)\theta = \mathrm{snd}(M\theta)$$

Note this postfix form, e.g. $M\theta$, for applying substitutions is traditional, in contrast to the prefix form, e.g. $f(M)$, you may be used to for functions. It is easy to see how to extend the above mapping if we extend the grammar of messages.

The **domain** of a substitution $\mathrm{dom}(\theta)$ is the variables that are changed by the substitution. That is, $\mathrm{dom}(\theta)$ is the subset of variables such that $x\theta \ne x$. For substitution θ, if $\mathrm{dom}(\theta) = \{x_1, x_2, \ldots x_n\}$, the **range** of a substitution, $\mathrm{ran}(\theta)$, is the set of messages $\{x_1\theta, x_2\theta, \ldots, x_n\theta\}$.

3.2 Symbolic reasoning on syntax in computer science

The concept of freshness can also be defined for a substitution θ, where $x \# \theta$ means that "$x \# \mathrm{dom}(\theta)$" and "$x \# \mathrm{ran}(\theta)$". The concept of freshness extends pointwise to sets of variables, terms and substitutions, for example $x, y \# \mathsf{M}, \mathsf{T}, \theta$ means both x and y are fresh for each of the terms and substitutions $M, \mathsf{T},$ and θ.

3.2.2.3 Avoiding capture of variables

Capture-avoiding substitutions are defined on terms which may contain bound variables, such as threads, in such a way that name clashes are avoided. This means that when we have situations such as $(\mathsf{new}\, x; \mathsf{out}(c, (x, y));0)\,\{y \mapsto \{x\}_k\}$, we must avoid the free variable x in message $\{x\}_k$ that y maps to from becoming confused with the local variable x that is bound by the fresh name binder. To do so, we must rename the bounded variable x with a fresh one and then we apply the substitution. In this case, if we rename x by z, we obtain $\mathsf{new}\, z; \mathsf{out}(c, (z, \{x\}_k));0$. Notice that the variable coming from the outer levels x has avoided being captured by the fresh name binder. Formally, we can define a capture-avoiding substitution on terms as follows, where $z \# \theta$, $\mathsf{new}\, x; \mathsf{T}$, $\mathsf{in}(\mathsf{M}, x); \mathsf{T}$, $\mathsf{let}\, x = M\, \mathsf{in}\, \mathsf{T}$.

$$0\theta = 0 \qquad (\mathrm{secret}(\mathsf{M}))\,\theta = \mathrm{secret}(\mathsf{M}\theta) \qquad \text{(Substitutions for threads)}$$

$$(\mathsf{out}(\mathsf{M}, \mathsf{N}); \mathsf{T})\theta = \mathsf{out}(\mathsf{M}\theta, \mathsf{N}\theta); \mathsf{T}\theta \qquad (\mathsf{in}(\mathsf{M}, x); \mathsf{T})\theta = \mathsf{in}(\mathsf{M}\theta, z); (\mathsf{T}\{x \mapsto z\}\theta)$$

$$(\mathsf{if}\, \mathsf{M} = \mathsf{N}\, \mathsf{then}\, \mathsf{T})\theta = \mathsf{if}\, \mathsf{M}\theta = \mathsf{N}\theta\, \mathsf{then}\, \mathsf{T}\theta \qquad (\mathsf{new}\, x; \mathsf{T})\theta = \mathsf{new}\, z; (\mathsf{T}\{x \mapsto z\}\theta)$$

$$(\mathsf{let}\, x = M\, \mathsf{in}\, \mathsf{T})\theta = \mathsf{let}\, z = M\theta\, \mathsf{in}\, \mathsf{T}\{x \mapsto z\}\,\theta$$

For example, for capture-avoiding substitution $(\mathsf{new}\, x; \mathrm{secret}(x))\,\{x \mapsto y\}$ results in a term $(\mathsf{new}\, z; \mathrm{secret}(z))$, where z is a fresh variable. Thus a substitution cannot accidentally change the meaning of bound variables, which is useful for modelling private information such as keys and nonces, which, recall, are typically random numbers that the attacker cannot guess in reasonable time (i.e., polynomial time with respect to some security parameter).

Notice that the capture-avoiding substitution for threads is no longer a function from threads to threads, since there are infinitely many possible choices of fresh name z. Recall, from you knowledge of discrete maths, that all functions should satisfy the property: if $x = y$ then $x\theta = y\theta$, i.e., any x maps to only one message $x\theta$. For this reason, we always assume that abstract syntax is quotiented by a relation that allows bound variables to be safely renamed; in other words, we make the assumption that in abstract syntax with binders we can always apply a notion called α-conversion, defined next.

A fundamental idea when working with syntax is the notion of α-conversion which allows the bound variables to be renamed. Programmers are used to the idea that when you write a program it does not matter what names you give to the local variables, and you can safety rename them without changing the behaviour of the program, as long as you avoid two distinct variables being given the same name.

To define α-conversion formally, firstly, recall that an *equivalence relation* is a reflexive, transitive, symmetric relation. A *congruence* is an equivalence relation that holds in any context. That is if T $\equiv_\alpha$ U then we have that the following also hold.

- in(M, x);T $\equiv_\alpha$ in(M, x);U
- out(M, N);T $\equiv_\alpha$ out(M, N);U
- if M = N then T $\equiv_\alpha$ if M = N then U

That is, a congruence relation allows you to replace one subterm with an equivalent subterm anywhere inside a term. Renaming or α-conversion can be defined as the least congruence $\equiv_\alpha$ such that the following hold, whenever $z \# T$.

$$\text{new } x; T \equiv_\alpha \text{new } z; (T\ \{x \mapsto z\}) \qquad \text{in}(M, x); T \equiv_\alpha \text{in}(M, z); (T\ \{x \mapsto z\})$$
(α-CONVERSION)

3.2.2.4 Composition and idempotency

Substitutions can be composed together to form new substitutions by composing their functions. Since substitutions are written using post fix notation this means that $x(\theta \circ \sigma) = (x\theta)\sigma$ (in contrast, recall functions, e.g. $\sin(x)$, are usually written as prefixes hence the order of composition is in reversed for substitutions compared to many mathematical texts and programming languages). For example, $\{x \mapsto \{y\}_k\} \circ \{y \mapsto (m, n)\}$ defines the substitution $\{x \mapsto \{(m, n)\}_k, y \mapsto (m, n)\}$. To see this, observe that $x(\{x \mapsto \{y\}_k\} \circ \{y \mapsto (m, n)\}) = \{y\}_k \{y \mapsto (m, n)\} = \{(m, n)\}_k$ and $y(\{x \mapsto \{y\}_k\} \circ \{y \mapsto (m, n)\}) = y \{y \mapsto (m, n)\} = (m, n)$, simply by applying the equation above to the previous definitions for substitutions.

A substitution is *idempotent* whenever $\theta \circ \theta = \theta$, which is equivalent to saying that for any variables $x, y \in \text{dom}(\theta)$ we have that $x \# y\theta$, that is, the messages in the range of the substitution are fresh for the domain.

3.2.3 Practice with the concepts of abstract syntax and substitutions

Exercise 3.1 Try these exercises to become familiar with the notions of freshness, α-conversion, and substitutions.

1. For thread $\text{out}(c, x); \text{new } x; \text{out}(c, x); 0$ indicate which occurrences of variables are free and which are binders and which are bound by a binder. Write down an α-equivalent term renaming the bound names so that they are all different from the free names.
2. Consider $\left(\text{new } k; \text{in}(c, x); \text{out}(c, \{(x, y)\}_k); 0\right) \{y \mapsto x\}$ with respect to capture-avoiding substitution. Write down a term for the thread resulting from applying

3.2 Symbolic reasoning on syntax in computer science   57

the above substitution to the above term. Ensure that the free variable x in the substitution is not captured by the input binder.
3. Pick a variable that is fresh for each of the above terms and another that is not fresh.
4. Indicate which of the following two functions are idempotent.

$$\{x \mapsto n, y \mapsto \{x\}_k, z \mapsto \{y\}_k\} \qquad \{x \mapsto n, y \mapsto \{n\}_k, z \mapsto \{\{n\}_k\}_k\}$$

5. Show how to obtain one of the above substitutions from the other using composition (twice).

3.2.4 Modelling threads in a Message Sequence Chart

Message sequence charts are intuitive diagrams widely used to describe protocols. Variants of message sequence charts appear throughout the security literature and technical specifications, in various engineering languages such as UML message sequence diagrams and, in are made precise in various verification techniques such as choreographies [97] and multiparty session types [66]. We use message sequence charts to indicate the roles of the protocol, the knowledge available to the roles, the messages that are sent and received, the order in which message occur, and also to convey security claims.

An example message sequence chart is presented in Fig. 3.4 where there are two roles i and r, the "initiator" and "responder" respectively, which appear in boxes at the top of threads. This example is taken from the paper which first introduced what has come to be known as the Dolev-Yao model, after the authors of that paper [48]. They presented this example of a protocol to illustrate that, even if we use perfect public-key encryption, then there is a way for an attacker to exploit the protocol in order to obtain a secret. We anticipate the fact that secrecy of message m fails by the crossed out security goal, where m is a nonce – a number generated for each session that is sufficiently large and random that the attacker cannot guess it without decrypting the message.

At first sight, it might not be immediately obvious in what sense the protocol fails to achieve the security goal of maintaining the secrecy of m whenever the initiator i completes a successful run of the protocol with an *honest* responder r. Indeed, if the agents running in the initiator or responder role only talk to each other, as was traditionally assumed to be the case in early enterprise architectures, it looks impossible to obtain the secret m, since it is only transmitted encrypted using the public key of the initiator or responder, hence can only be decrypted by the initiator or responder, who, when they play honestly by the rules of the protocol, will never reveal their private keys sk_i and sk_r. For this reason, we must make our trust assumptions clear: we assume that, in addition to two honest agents talking there may be concurrent sessions with other agents that may or may not be honest.

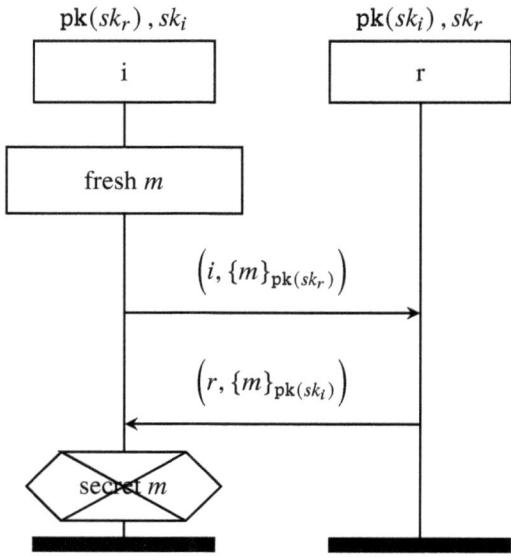

Fig. 3.4 Dolev-Yao 1: a broken unilateral secret-sharing protocol.

Making our assumptions for our secrecy claim precise requires formal machinery. The first step towards that aim is to model the threads corresponding to roles i and r in the MSC diagram using our syntax for threads.

The initiator in this protocol we model using the thread below. We define an abbreviation for the thread *Initiator*, which is parameterised on the free variables c, i, sk_i, r, that appear in the term used to model the thread.

Initiator$(c, i, sk_i, r, pk_r) \triangleq$
new m;   Declare a fresh nonce m.
out$\left(c, \left(i, \{m\}_{pk_r}\right)\right)$;   Send a message to a responder.
in(c, x);   Receive a response from the network.
if fst$(x) = r$ then   Check the message is from the same responder.
if dec(snd(x), sk_i) = m then   Check that the message received back is m.
secret(m) Make the assertion that m is secret.

All messages in the thread are sent or received on the channel c, which will be made public so that the messages can be intercepted by the network, which may be influenced by the attacker. The identity of the channel is not too important for secrecy, but will play a role in more advanced properties; hence we include channels even though the channel on which inputs and outputs occur is not made explicit in some

3.2 Symbolic reasoning on syntax in computer science

protocol modelling languages such as Scyther [47], or in the message sequence chart for this protocol. Channels are however made explicit in tools such as ProVerif [28] and DeepSec [41].

The responder in this protocol is modelled as follows.

$Responder(c, r, sk_r, i, pk_i) \triangleq$
    $\text{in}(c, x);$ Receive a message from the network.
    if $\text{fst}(x) = i$ then   Check that the message comes from the initiator.
    let $m = \text{dec}(\text{snd}(x), sk_r)$ in   Decrypt the secret using private key sk_r.
    $\text{out}\left(c, \left(r, \{m\}_{pk_i}\right)\right)$ Send m to the initiator using their public key.

We do not make any assertions that m is secret from the perspective of the responder, reflecting the fact that the responder does not aim to achieve this secrecy property in this protocol. To see why, observe that we assume the identity i and the public key pk_r are public, that is, they are known to any attacker. Hence an attacker may send a message of the form $\left(i, \{known\}_{pk_r}\right)$ and the responder r would react anyway. The (claimed) objective of this Dolev-Yao 1 protocol is what is called *unilateral secrecy*, in the sense that secrecy should hold only for one role of the protocol; in this case the secrecy of m holds only from the perspective of the initiator.

At first reading, this may be surprising that a message is not simply secret or not secret – secrecy is dependent on the perspective. Even if unilateral secrecy seems weak for many security protocols, there may be scenarios where only the initiator (e.g., a terminal) must know that a specific agent has received a secret token, before exchanging resources by making use of that token. Since the initiator knows that the channel the token is secret, the initiator may trust information received from the responder encrypted using that token, but the responder may not trust information received from the initiator, since the initiator does not have the same guarantees. If the desired secrecy property were to hold, an initiator can assume that only a specific responder received the desired message and, if honest, will keep it confidential. Unilateral security properties are not esoteric: e.g., TLS only ensures the secrecy of information received from a trusted server to the client, but the server has no guarantees about the client. However, as we will show in detail in Chapter 4, no secrecy property holds for the Dolev-Yao 1 protocol, so even unilateral secrecy fails.

3.2.5 Practice in modelling protocol roles as threads

For many protocols secrecy is a multilateral property that should hold for both the initiator and responder. Some protocols employ techniques to maintain the secrecy of messages that are received by the initiator. In the case of the Needham-Schroeder protocol presented in the message sequence chart in Fig. 3.5, the use of three messages to create a round-trip for both agents is intended to maintain secrecy goals for the responder.

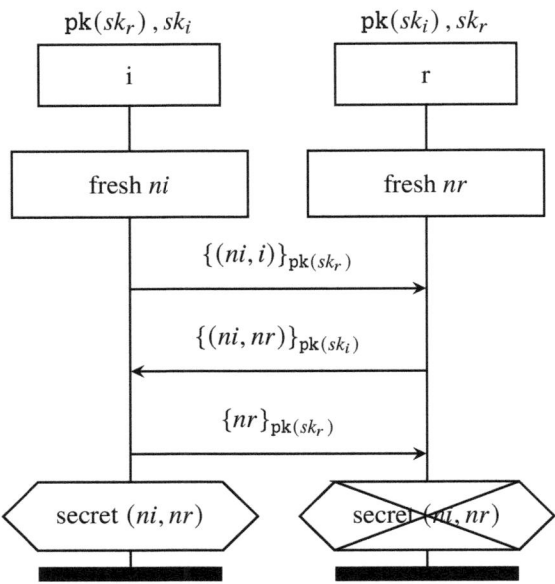

Fig. 3.5 The Needham-Schroeder protocol.

An original security goals of the Needham-Schroeder protocol was to establish shared secrets *ni* and *nr*. We will see later that the Needham-Schroeder protocol failed to meet its own security requirements, as hinted by the fact that one of the two assertions is crossed out. However, to see this we will require more formal machinery and to appreciate the story behind the need for evolving security requirements, which continues to this day.

Exercise 3.2 Write down terms modelling the initiator and responder threads for the Needham-Schroeder protocol. In this protocol, include a secrecy claim for both the initiator and responder. As with threads of the Dolev-Yao 1 protocol, we include an explicit channel *c* for all inputs and outputs for each thread.

3.2.6 A digression into music and back to security protocols

Readers who have played music to a reasonable level may grasp the following analogy (or sport or any other activity to a decent level). If all you do is use your intuition and try to copy sounds and words, then the best you will ever be is a karaoke singer. In order to master music, you must learn and practice your scales, you must be able to steadily measure your rhythm, understand harmony; and once you have practised these core elements you can move on to techniques such as counterpoint, resolution and timbre; thereby opening up paths for expressing yourself which you were never able to see without studying music. Of course, a musician still needs emotional intelligence and creative inspiration, but such qualities cannot reach their potential until we are confident with assimilating the basic elements. Projecting this view back to the world of security protocols, we learn that it is important that we understand carefully and practice the rules of the systems introduced before we jump to conclusions about whether a security goal is achieved. For this reason, we begin with the basics of syntax and semantics and include exercises to encourage the reader to practice the rules of the game.

For ambitious readers: observe that, once the rules are mastered, a musician may find ways to creatively break rules of harmony and fuse styles to create new sounds, or tackle intricate compositions; much like a researcher creates new models to capture emerging security requirements and protocols. However, the same basic rules that have been honed by decades of research are constantly appealed to, even when conducting bleeding edge research.

3.3 Modelling protocol behaviour using operational semantics

So far, we provided syntactic means for modelling protocol behaviour. For their meaning, we provided intuition. Now, we will introduce more formal treatment by specifying a semantics.

In the early days of computer science, when defining a programming language or system, a reference implementation would be provided for this. A compiler engineer would determine whether their own implementation was correct by running their program next to the reference implementation to test whether the behaviours matched. Beginning with ideas developed for making precise languages such as ALGOL 60, functional programming languages such as LISP, and logic programming languages such as Prolog, the field of structural operational semantics emerged [98]. Structural operational semantics provides a more precise methodology for designing and verifying systems where we define an executable model, directly on the syntax of the language or system. This enables us to not only give a more precise reference for implementations, but also provides us with ways for verifying static compile-time and run-time properties such as type safety, memory safety, deadlock freedom, or compliance to functional specifications.

Today we have operational semantics for many programming languages, e.g., the operational semantics for C is more complete than any reference implementation [64]. There are operational semantics that have guided the development of type systems for Java [72], there are operational semantics for diverse systems such as smart contracts [83], secure micro kernels [77], and networked systems [74, 100] – the literature on operational semantics is vast. Some systems take verification using operational semantics to the extreme and verify system stacks right down to their underlying hardware [79].

Security is a key domain in which operational semantics is useful, since security flaws are usually due to things that we cannot immediately see using solely the raw intuition of experts. Using operational semantics, we can precisely specify the behaviour of a protocol and check that all behaviours conform with security and privacy properties. We will model the system at an appropriate level of abstraction, where we are concerned with the interactions between the *honest* agents running the protocol and a network on which we cannot assume that everyone is trusted.

3.3.1 An equational approach to defining functions on messages

When dealing with messages, $\mathsf{dec}(\{M\}_K, K)$ and M are syntactically different but semantically, they should mean the same. Such operational behaviour of functions can be specified in terms of equations, which define the effect of applying *deconstructors* such as decryption to *constructors* such as encryption. A constructor builds a message, whereas a deconstructor is a function that pulls it apart. We make use of an equational theory with the following equations.

$$\begin{aligned} \mathsf{fst}((M, N)) &=_E M & \text{(First)} \\ \mathsf{snd}((M, N)) &=_E N & \text{(Second)} \\ \mathsf{dec}\left(\{M\}_{\mathsf{pk}(K)}, K\right) &=_E M & \text{(Decrypt)} \end{aligned}$$

Like α-conversion, the equational theory defines a congruence, which allows the rules to be applied anywhere inside a message term that matches the pattern in the rule. For example, the following equalities follow from the above rules.

$$\mathsf{fst}\left(\mathsf{dec}\left(\{(m, n)\}_{\mathsf{pk}(sk_a)}, sk_a\right)\right) =_E \mathsf{fst}((m, n)) =_E m$$

The above example allows a message encrypted with public key $\mathsf{pk}(sk_a)$ to be decrypted using the corresponding private key sk_a, and furthermore allows us to extract the first message. We name explicitly the equational theory $=_E$, since, later, in order to model some protocols, we will require additional cryptographic features which require more equations or different equations, e.g., for symmetric encryption, advanced cryptographic features, or even more controlled variants of public key encryption.

Remark 3.2 Some tools make use of a simple type system on messages to prevent meaningless terms occurring. For example $\mathsf{fst}\left(\{x\}_y\right)$ is meaningless, since the first projection deconstructor should only be applied to pairs. In the theory in this work, a type system is optional; hence we do not impose a typing discipline. None-the-less types will appear in some examples of code for tools, since types generally help to

3.3 Modelling protocol behaviour using operational semantics

encourage better coding. While a type system has a precise definition, we do not provide it here, and leave that aspect to the intuition of the coder.

3.3.2 States for labelled transitions

In order to define a structural operational semantics we use a syntax for representing system states and a transition relation defining how we move from one state to the next, much like an interpreter for a programming language. For interactive systems, such as protocols where there are inputs and outputs with the environment as well as internal state changes, we make use of *labelled transitions* where the label on a transition indicates the action that is performed, such as an input or an output.

To specify a protocol as a Labelled Transition System (LTS) we require:

- Syntax: how to write down states through which a system passes.
- Operational semantics: how to know which transitions between states are allowed.

3.3.2.1 A syntax for states of a protocol

The **states** of the system record both the honest processes currently running, and also record the knowledge that is made available to the environment by outputting messages. In a state we allow substitutions, known as *active substitutions* and denoted θ below, to float alongside threads (or more general processes when we extend this model). This active substitution and process appear in the scope of zero or more new name binders, as defined by the following grammar.

$State_A$ ::= $[\theta \mid P]$ An active substitution and a thread (or process).
         | new x; $State_A$ Fresh name binders hiding private information.

(SYNTAX OF STATES)

According the above grammar, if P is a process such as a thread T and θ is a substitution, then an *extended process* is of the form "new x_1; new x_2; ... new x_n; $[\theta \mid P]$". Such multiple fresh name binders can be abbreviated as "new $x_1, x_2, \ldots x_n$; $[\theta \mid P]$", by listing all the fresh names together. The more general term process encompasses not only threads, but also richer processes that allow multiple threads to run concurrently. In the literature, these states are sometimes referred to as *extended processes*, since they are effectively processes extended with an active substitution.

The active substitution represents state information, recording the knowledge available to an attacker. You can regard the substitution as a database of all the messages that have been output to the network and intercepted by the attacker. For example, consider the following state.

$$\text{State}_1 \triangleq \text{new } sk_a, m; \left[\left\{ \begin{array}{l} x \mapsto \text{pk}(sk_a), \\ y \mapsto \{m\}_{\text{pk}(sk_e)} \end{array} \right\} \middle| \begin{array}{l} \text{in}(c, z); \\ \text{if } \text{dec}(z, sk_a) = m \text{ then} \\ \text{secret}(m) \end{array} \right] \quad (\text{State } 1)$$

The above state indicates that the attacker has intercepted two outputs x and y, which are aliases for the public key $\text{pk}(sk_a)$ and the ciphertext $\{m\}_{\text{pk}(sk_e)}$ respectively. The attacker however cannot directly see the secret key sk_a or the message m, which are hidden by the fresh binder, although it knows about free variables such as sk_e. The thread, is ready to receive a message from the environment which it will bind to z and check whether the message is m encrypted in such a way that only the owner of sk_a can read it. The thread ends by asserting that m is secret, which we will see below is not the case. We will use this state as a running example to explain the rules of the system.

We assume that extended processes are in a **normal form**, that is, if $\text{State}_1 = \text{new } x_1, \ldots, x_n; [\theta \,|\, P]$, then θ is idempotent and $\text{dom}(\theta)$ # P and $x_1, \ldots, x_n$. In other words, we assume that the substitution θ has been applied to the process P, since there are no such variables; and thus the variables in the domain of the substitution are used only as aliases for referring to previously output messages from the perspective of the attacker. We can see that the example (STATE 1) above satisfies these conditions.

In summary, a state can be of the following form.

$$\text{new } \mathbf{x}; [\theta \,|\, P] \triangleq \text{new } \underbrace{x_1, \ldots, x_k}_{\text{nonces \& keys}}; \left[\left\{ \underbrace{u_1, \ldots, u_n}_{\text{aliases, } domain} \mapsto \overbrace{M_1, \ldots, M_n}^{\text{messages, } range} \right\} \middle| P \right]$$

In the above, $\mathbf{x}$ lists variables that represents private information (private keys, nonces) generated so far. Active substitution θ lists labelled messages sent by honest agents to the network so far. A process P represents future, not yet executed, actions.

3.3.3 The labelled transition system: inference rules

We define the labelled transition system using a set of *inference rules*. Inference rules are written in the following form.

$$\frac{\text{Premise 1} \quad \cdots \quad \text{Premise } n}{\text{Conclusion}} \text{ (rule name)}$$

An inference rule is such that if all the premises above a horizontal line hold, then the conclusion below the line holds. We explain next the labels and inference rules for threads presented in Fig. 3.6. Each construct for processes and states has an inference rule. Those inference rules are combined into a tree of rules instances, where the premise of one rule is the conclusion of another rule.

3.3 Modelling protocol behaviour using operational semantics

$$\frac{\dfrac{\overline{\text{Axiom 1}}\;(\text{axiom 1})}{\text{Claim 1}}\;(\text{rule 3}) \qquad \dfrac{\overline{\text{Axiom 2}}\;(\text{axiom 2})}{\text{Claim 2}} \qquad \dfrac{\overline{\text{Axiom 3}}\;(\text{axiom 3})}{\;}\;(\text{rule 3})}{\text{Conclusion}}\;(\text{rule 1})$$

We will use such "derivation trees" in the following to calculate input and output actions and also the effect of the input or output action on the next state. The leaves of a derivation tree must be axioms, which are rules without premises. Examples of such trees appear in the following subsections.

3.3.3.1 The labels

In a labelled transition of the form $\text{State}_A \xrightarrow{\pi} \text{State}_B$ a label π decorates the transition relation. The labels represent what the attacker can see when they interact with the protocol through the network while the honest agents move from one state to another. Labels are defined using the following grammar.

$\pi ::=$ out(M, x) Intercept message sent, on channel M, to be referred to as x.
    $\mid$ in(M, N) Force the message N to be received, on channel M.
    $\mid$ secret(M) Assertion that attacker can see a secret by using message M.

(SYNTAX OF LABELS)

An attacker can observe outputs, e.g., out(c, x) which represents the interception of an output on channel c where the message intercepted can be referred to in the future using variable x, which is an alias for a message. The attacker may also observe inputs which she may influence by modifying or inserting the input message, e.g., the input in$(c, \text{dec}(x, y))$ represents an input on channel c which is obtained by decrypting the message represented by x using the a private key represented by y, where x and y may be previously intercepted messages. The attacker may also declare that they know a secret, that should be hidden from them using the label secret$(\text{fst}(\text{dec}(x, k)))$, which says that we can produce a message that should have been secret message by decrypting the message x using key k and then taking the first element of the pair of messages obtained from the deciphered message.

We use the symbol π for labels. This respects a tradition where π stands for "prefix" in the sense that the input and output labels are also prefixes in the syntax of threads. This tradition dates back to the π-calculus [92] on which the model we are introducing is based.

Next we analyse in detail the interpretation of each of the rules in Fig. 3.6.

3.3.3.2 The ALIAS rule

Transitions are designed according to the principle that the attacker is an observer interacting with the actions on the labels and cannot directly observe the internal

$$\frac{\mathsf{T}_0 \xrightarrow{\pi\theta} \mathsf{new}\,\mathbf{y};[\sigma\,|\,\mathsf{T}_1] \quad \mathbf{x}\,\#\,\pi \quad \mathsf{dom}(\sigma),\mathbf{y}\,\#\,\theta}{\mathsf{new}\,\mathbf{x};[\theta\,|\,\mathsf{T}_0] \xrightarrow{\pi} \mathsf{new}\,\mathbf{x},\mathbf{y};[\theta\circ\sigma\,|\,\mathsf{T}_1]} \quad (\text{Alias})$$

$$\frac{\mathsf{K} =_\mathsf{E} \mathsf{M} \quad x\,\#\,\mathsf{K},\mathsf{M},\mathsf{N},\mathsf{T}}{\mathsf{out}(\mathsf{M},\mathsf{N});\mathsf{T} \xrightarrow{\mathsf{out}(\mathsf{K},x)} [\{x \mapsto \mathsf{N}\}\,|\,\mathsf{T}]} \quad (\text{Output})$$

$$\frac{\mathsf{K} =_\mathsf{E} \mathsf{M}}{\mathsf{in}(\mathsf{M},x);\mathsf{T} \xrightarrow{\mathsf{in}(\mathsf{K},\mathsf{N})} [id\,|\,\mathsf{T}\,\{x \mapsto \mathsf{N}\}]} \quad (\text{Input})$$

$$\frac{\mathsf{T}\,\{x \mapsto z\} \xrightarrow{\pi} \mathsf{State} \quad z\,\#\,\pi,\mathsf{new}\,x;\mathsf{T}}{\mathsf{new}\,x;\mathsf{T} \xrightarrow{\pi} \mathsf{new}\,z;\mathsf{State}} \quad (\text{Extrude})$$

$$\frac{\mathsf{T}_1 \xrightarrow{\pi} \mathsf{State} \quad \mathsf{M} =_\mathsf{E} \mathsf{N}}{\text{if } \mathsf{M}=\mathsf{N} \text{ then } \mathsf{T}_1 \text{ else } \mathsf{T}_0 \xrightarrow{\pi} \mathsf{State}} \quad (\text{Match}) \qquad \frac{\mathsf{T}_0 \xrightarrow{\pi} \mathsf{State} \quad \mathsf{M} \neq_\mathsf{E} \mathsf{N}}{\text{if } \mathsf{M}=\mathsf{N} \text{ then } \mathsf{T}_1 \text{ else } \mathsf{T}_0 \xrightarrow{\pi} \mathsf{State}} \quad (\text{Mismatch})$$

$$\frac{\mathsf{N} =_\mathsf{E} \mathsf{M}}{\mathsf{secret}(\mathsf{M}) \xrightarrow{\mathsf{secret}(\mathsf{N})} [id\,|\,0]} \quad (\text{Secret}) \qquad \frac{\mathsf{T}\,\{x \mapsto \mathsf{M}\} \xrightarrow{\pi} \mathsf{State}}{\mathsf{let}\,x=\mathsf{M}\,\mathsf{in}\,\mathsf{T} \xrightarrow{\pi} \mathsf{State}} \quad (\text{Let})$$

Fig. 3.6 Rules of a labelled transition system for threads, and for the assertion secret.

state. For example, they cannot directly view messages involving private variables hidden using the fresh binders, unless there is a way for the attacker to deduce a message using the information intercepted on the network, that is, using the aliases in the domain of the active substitution of the current state.

This constraint on the capabilities of the attacker is enforced by the ALIAS rule which does two things. Firstly, the ALIAS rule prevents such private information from appearing in the label π, by using a constraint $\mathbf{x}\,\#\,\pi$, where $\mathbf{x}$ is a set of names bound at the outermost level of the extended process representing the state. In the definition of freshness we assume all variables appearing in π are free. To be explicit we assume the following.

$$\mathsf{fv}(\mathsf{secret}(\mathsf{M})) = \mathsf{fv}(\mathsf{M}) \qquad (\text{Free variables of labels})$$

$$\mathsf{fv}(\mathsf{out}(\mathsf{M},x)) = \mathsf{fv}(\mathsf{M}) \cup \{x\} \qquad \mathsf{fv}(\mathsf{in}(\mathsf{M},\mathsf{N})) = \mathsf{fv}(\mathsf{M}) \cup \mathsf{fv}(\mathsf{N})$$

Secondly, the ALIAS rule applies the current active substitution to the label in the premise of the rule. This means that any aliases appearing on the label are instantiated

3.3 Modelling protocol behaviour using operational semantics 67

with their corresponding messages. This is useful for several other rules that compare a message appearing on the label with a message appearing in the thread, in order to determine whether the rule is enabled.

Next, we provide examples of how the ALIAS rule works with other rules in order to constrain realistically the power of an attacker when building messages in order to interact with honest threads (recall all threads we model explicitly are assumed to be honest, since they follow the protocol).

3.3.3.3 The OUTPUT rule

Since our assumption is that the attacker is able to control the network, messages that are sent to the network can be intercepted by the attacker. This assumption is part of what is known as the Dolev-Yao model. The OUTPUT rule says that when an output occurs we allocate it a fresh name, which can be used as an alias to refer that particular output that has been intercepted and we extend the active substitution appropriately. This has the effect of updating the database of knowledge of the attacker whom intercepts outputs and records them for future use.

For example, consider the following instance of the OUTPUT rule, where a public key is sent on a channel named *keys*.

$$\cfrac{keys =_E keys \quad pk_a \mathbin{\#} keys, \mathrm{pk}(sk_a) \, , \; \begin{array}{l} \mathsf{new}\, m; \\ \mathsf{out}\!\left(c, \{m\}_{\mathrm{pk}(sk_e)}\right); \\ \mathsf{in}(c, z); \\ \mathsf{if}\, \mathrm{dec}(z, sk_a) = m\, \mathsf{then} \\ \mathsf{secret}(m) \end{array}}{\begin{array}{l} \mathsf{out}(keys, \mathrm{pk}(sk_a)); \\ \mathsf{new}\, m; \\ \mathsf{out}\!\left(c, \{m\}_{\mathrm{pk}(sk_e)}\right); \\ \mathsf{in}(c, z); \\ \mathsf{if}\, \mathrm{dec}(z, sk_a) = m\, \mathsf{then} \\ \mathsf{secret}(m) \end{array} \xrightarrow{\mathsf{out}(keys, pk_a);} \left[\begin{array}{l} \{pk_a \mapsto \mathrm{pk}(sk_a)\} \\ \mathsf{new}\, m; \\ \mathsf{out}\!\left(c, \{m\}_{\mathrm{pk}(sk_e)}\right); \\ \mathsf{in}(c, z); \\ \mathsf{if}\, \mathrm{dec}(z, sk_a) = m\, \mathsf{then} \\ \mathsf{secret}(m) \end{array}\right]}\;\mathsf{OUTPUT}$$

In the above we have taken the liberty to write the extended process on the right of the transition vertically, which we will do so sometimes throughout this book simply to save space as expressions can be large. Notice that the variable pk_a is chosen to be fresh with respect to all messages and threads that appear. Any other fresh variable could have been chosen; the name provided here is just to help remember that that pk_a is an alias for a message representing a public key $\mathrm{pk}(sk_a)$. The freshness conditions ensure that the variable chosen is a unique alias for that particular output.

Applying the ALIAS rule then ensures that pk_a is fresh also with respect to other aliases and messages in the context. We assume in this case that the context is the identity substitution *id* only, so there are no additional checks. After applying the ALIAS rule we obtain a derivation tree with the following transition from extended processes to extended processes as its conclusion.

$$
id \begin{bmatrix} \begin{vmatrix} \text{out}(keys, \text{pk}(sk_a)); \\ \text{new } m; \\ \text{out}\Big(c, \{m\}_{\text{pk}(sk_e)}\Big); \\ \text{in}(c, z); \\ \text{if } \text{dec}(z, sk_a) = m \text{ then} \\ \text{secret}(m) \end{vmatrix} \end{bmatrix} \xrightarrow{\text{out}(keys, pk_a);} \begin{bmatrix} \dfrac{\{pk_a \mapsto \text{pk}(sk_a)\}}{\text{new } m;} \\ \text{out}\Big(c, \{m\}_{\text{pk}(sk_e)}\Big); \\ \text{in}(c, z); \\ \text{if } \text{dec}(z, sk_a) = m \text{ then} \\ \text{secret}(m) \end{bmatrix}
$$

This output transition by itself is not particularly useful for modelling security protocols. Usually a cryptographic message sent involves a secret, such as a private key or nonce. In the above transition, notice that the key sk_a is sent as a free variable – it appears free in the threads and messages and is not even bound by a fresh name quantifier in the extended process. Since sk_a is free, this means that it is known to the attacker. The EXTRUDE rule, explained next shows how to perform a similar output, modified such that the private key sk_a remains secret. This will help the reader see one reason why we do not put the message $\text{pk}(sk_a)$ on the output labels. There are many reasons for having aliases for the message output on output labels, for which we refer the curious reader to the considerable body of work on the theory of the applied π-calculus [8, 70, 15].

3.3.3.4 The EXTRUDE rule

The Oxford dictionary definition of extrude is, "extrude (something) (from something) (formal) to force or push something out of something; to be forced or pushed in this way". Extrusion in the context of process calculi such as the π-calculus is the ability to push out names [92, 91]. By pushing out names, the scope in which they are declared enlarges to encompass a larger context in which they may be referred to as they are exchanged through a network. For example, if a name representing a nonce appears in a message that is being output then the message is intercepted by the attacker, as represented by the active substitution in an extended process, and hence the scope of the name must encompass the active substitution in the extended process. The scope of the name is pushed out; that is, the name is extruded.

We can see extrusion at play by adapting the output transition above. In what follows, the secret key sk_a is bound by a fresh name binder in the process on the left of the labelled transition in the conclusions of the derivation tree. Notice that sk_a remains bound after the transition, but at the outermost level of an extended process.

3.3 Modelling protocol behaviour using operational semantics

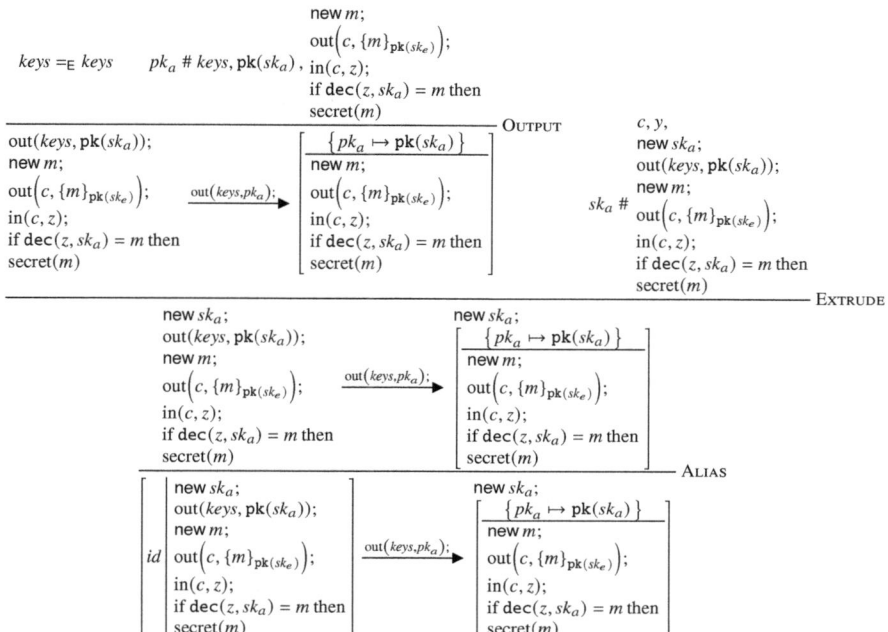

In the above, the OUTPUT rule is exactly as seen previously. The EXTRUDE has side conditions that simply check that sk_a does not appear on the label, since the label can only refer to knowledge of the attacker and the attacker does not a priori know sk_a. That bookkeeping is an essential part of the separation between the state of honest participants and the labels representing interactions with attackers: bound names may appear directly in threads representing honest participants; while attackers may only refer to such secrets indirectly via aliases for messages in which the name of the secret appears. In this case, for example, there is no way for the attacker to refer even indirectly to the sk_a, thus it may only use the public key $\mathrm{pk}(sk_a)$ as a whole.

As seen previously when discussing the OUTPUT rule, the ALIAS rule is then applied in a trivial way. Notice in the conclusion of the tree above, the ALIAS rule has lifted a transition from a thread to an extended process to one between extended process only. Since extended process are the states of our calculus we therefore have derived a labelled transition between states by using the rules of the operational semantics.

We go though another example of extrusion in detail, since it is a powerful concept for modelling secrets in security protocols, central to the methodology in this book. Consider the following output transition, where the state on the right of the labelled transition is the example state (STATE 1) introduced previously.

$$\begin{bmatrix} \text{new } sk_a; \\ \dfrac{\{x \mapsto \text{pk}(sk_a)\}}{\text{new } m;} \\ \text{out}\big(c, \{m\}_{\text{pk}(sk_e)}\big); \\ \text{in}(c, z); \\ \text{if } \text{dec}(z, sk_a) = m \text{ then} \\ \text{secret}(m) \end{bmatrix} \xrightarrow{\text{out}(c,y)} \begin{bmatrix} \text{new } sk_a, m; \\ \dfrac{\left\{\begin{array}{l} x \mapsto \text{pk}(sk_a), \\ y \mapsto \{m\}_{\text{pk}(sk_e)} \end{array}\right\}}{\text{in}(c, z);} \\ \text{if } \text{dec}(z, sk_a) = m \text{ then} \\ \text{secret}(m) \end{bmatrix}$$

(TRANSITION 1)

The above labelled transition is enabled by three inference rules EXTRUDE, ALIAS and OUTPUT as shown in Fig. 3.7.

Firstly, we apply the ALIAS rule in order to look at the thread inside the extended process on the left of the transition. This removes the fresh name quantifier binding sk_a at the outermost level and checks that sk_a does not appear on the label. Secondly, we apply the EXTRUDE rule to extrude the scope of the fresh variable m to the outermost level of the extended process to the right of the transition. Thirdly, we apply the OUTPUT rule by selecting a fresh name y which is used as an alias for the output message $\{m\}_{\text{pk}(sk_e)}$. The ALIAS rule also ensures that y is fresh with respect to the extended process, and, furthermore, updates the active substitution by composing substitutions as follows.

$$\{x \mapsto \text{pk}(sk_a)\} \circ \{y \mapsto \{m\}_{\text{pk}(sk_e)}\} = \{x \mapsto \text{pk}(sk_a),\ y \mapsto \{m\}_{\text{pk}(sk_e)}\}$$

This way the active substitution of STATE 1, recalled above, has been extended with the mapping reflecting the output messages and choice of alias.

3.3.3.5 The rule INPUT

To understand the input rule consider again the state STATE 1, in which the thread is ready to receive something on channel c. That state can perform many input transitions by varying the message on the input label. In the following input transition, notice that the label $\{\text{dec}(y, sk_e)\}_x$ appears.

$$\text{State}_1 \xrightarrow{\text{in}(c, \{\text{dec}(y, sk_e)\}_x)} \begin{bmatrix} \text{new } sk_a, m; \\ \dfrac{\left\{\begin{array}{l} x \mapsto \text{pk}(sk_a), \\ y \mapsto \{m\}_{\text{pk}(sk_e)} \end{array}\right\}}{\text{if } \text{dec}\left(\left\{\text{dec}\big(\{m\}_{\text{pk}(sk_e)}, sk_e\big)\right\}_{\text{pk}(sk_a)}, sk_a\right) = m \text{ then}} \\ \text{secret}(m) \end{bmatrix}$$

(TRANSITION 2)

This transition makes use of the ALIAS rule and INPUT rules as shown in the first derivation presented in Fig. 3.7. In the derivation in Fig. 3.7, firstly, we apply the ALIAS rule, which checks that bound variables sk_a and m do not appear on the label, which represent the message from the perspective of the attacker. Observe that the

3.3 Modelling protocol behaviour using operational semantics

message on the input label $\{\text{dec}(y, sk_e)\}_x$ refers to variables x and y in the domain of the substitution; hence it means, "decrypt the output that was intercepted with alias y using the secret key of the attacker sk_e, then encrypt it again using the other output that was intercepted with alias x.

To move from the view of the attacker to the view of the honest agent, who knows the secrets sk_a and m we simply apply the active substitution to the message on the label. This is achieved in the premise of the ALIAS rule, where the active substitution is applied to the label to expand the aliases x and y. That is, we perform the following evaluation.

$$\{\text{dec}(y, sk_e)\}_x \{x \mapsto \text{pk}(sk_a), y \mapsto \{m\}_{\text{pk}(sk_e)}\} = \left\{\text{dec}\left(\{m\}_{\text{pk}(sk_e)}, sk_e\right)\right\}_{\text{pk}(sk_a)}$$

Next we apply the INPUT rule which removes the receive prefix from the process, advancing the state, and substitutes all occurrences of the input variable z with the message above, obtained from the input label. Notice that we cannot place directly the message $\{m\}_{\text{pk}(sk_a)}$ on the label in the conclusion of the derivation above since that would imply that the attacker has the power to directly refer to m and sk_a which are hidden by the fresh binder.

3.3.3.6 The MATCH rule

We continue the running example above. Consider the state that we reach after (TRANSITION 2) above and let us call the resulting state State_2. In State_2, there is an if condition that can only proceed if the guard

$$\text{``dec}\left(\left\{\text{dec}\left(\{m\}_{\text{pk}(sk_e)}, sk_e\right)\right\}_{\text{pk}(sk_a)}, sk_a\right) = m\text{''}$$

holds in the if-statement. The guard does indeed hold, since we have, according our equational theory, we have the following.

$$\text{dec}\left(\left\{\text{dec}\left(\{m\}_{\text{pk}(sk_e)}, sk_e\right)\right\}_{\text{pk}(sk_a)}, sk_a\right) =_E \text{dec}\left(\{m\}_{\text{pk}(sk_a)}, sk_a\right) =_E m$$

Thus we can apply the MATCH rule. However, the match rule does not, by itself, produce any actions. We must look at the next prefix in the thread in order to know what to do, which happens to be the special assertion $secret(m)$ which we explain below.

Note that the MISMATCH rule is similar to MATCH but triggers and else branch if the messages cannot be made equal. Mismatch will play a role in the next chapter when defining threat models.

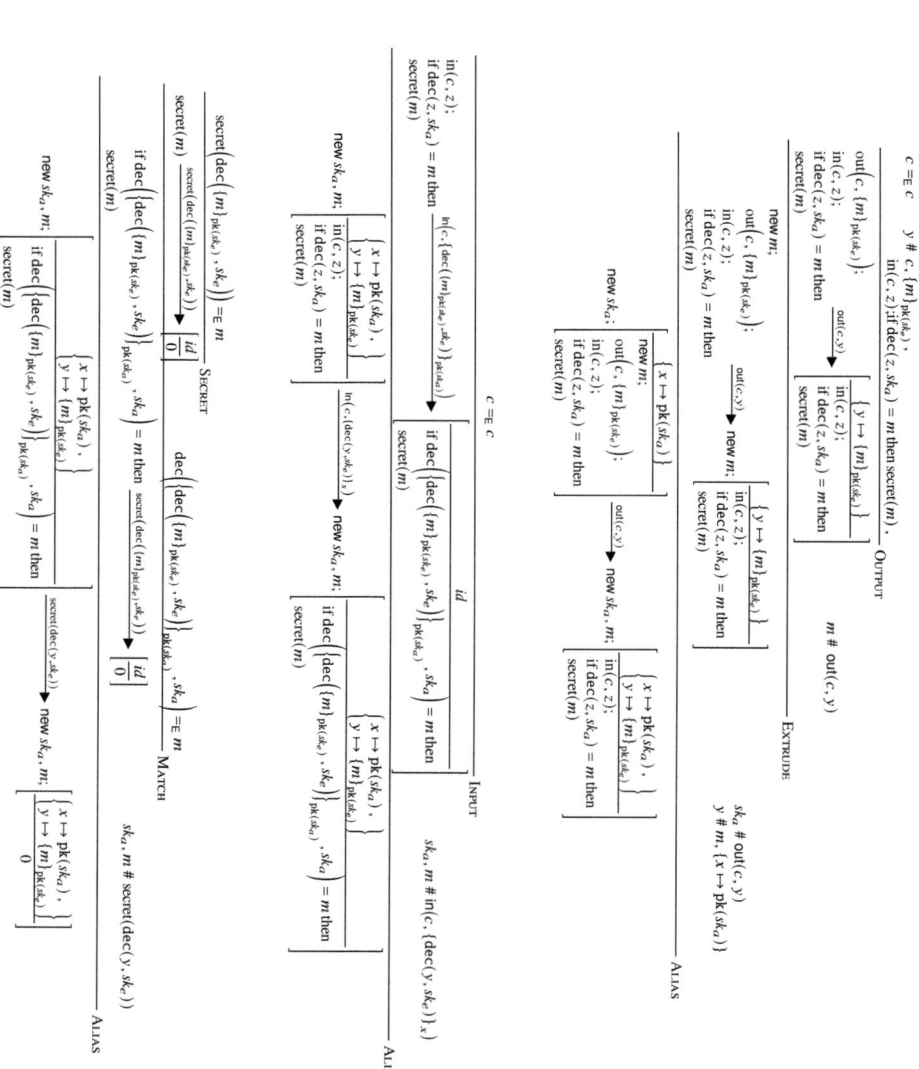

Fig. 3.7 The derivations for the three transitions of our running example.

3.3 Modelling protocol behaviour using operational semantics 73

3.3.3.7 The SECRET rule

The SECRET rule, in Fig. 3.6, is really part of a formulation of the security property *secrecy*, to be defined in the next section, rather than a behaviour of threads. The idea is that the rule can never occur if the message is really secret. In our running example, the rule is used to enable the following transition, in conjunction with the MATCH rule, as described above, and the ALIAS rule that forbids us from referring directly to m and sk_a in the message on the label.

$$\text{State}_2 \xrightarrow{\text{secret}(\text{dec}(y, sk_e))} \text{new } sk_a, m; \left[\left\{ \begin{array}{l} x \mapsto \text{pk}(sk_a), \\ y \mapsto \{m\}_{\text{pk}(sk_e)} \end{array} \right\} \middle| 0 \right] \quad \text{(TRANSITION 3)}$$

This labelled transition can be derived as shown in the Fig. 3.7.

Similarly to other rules, the message $\text{dec}(y, sk_e)$ is the view of the attacker of the secret: "decrypt the message intercepted and assigned alias x, using the secret key of the attacker sk_e, which is a free variable." Indeed, if we apply the current active substitution to the message on the label we obtain m by using the following evaluation.

$$\text{dec}(y, sk_e) \left\{ x \mapsto \text{pk}(sk_a), y \mapsto \{m\}_{\text{pk}(sk_e)} \right\} = \text{dec}\left(\{m\}_{\text{pk}(sk_e)}, sk_e \right) =_{\mathsf{E}} m$$

Notice in the derivation in Fig. 3.7 achieves the above effect through the ALIAS rule applying the current active substitution to the label and the SECRET rule checking that the messages are equivalent according to our equational theory. The existence of this transition labelled with a secret action shows that the attacker can control the network in such a way that secrecy of m is violated. More specifically, once the honest thread modelled above runs to completion, according to the transitions we have described, then the attacker can tiger the secret action that should only be possible if they know the secret.

Now consider a contrasting example, where a secret is not revealed to the attacker. In the thread below a message is sent encrypted using the public key of an honest agent $\text{pk}(sk_b)$, and that the attacker does not know the corresponding secret key sk_b. For this scenario, we begin with the following output transition.

$$\begin{array}{c} \text{new } sk_b; \\ \left[\begin{array}{c} \{x \mapsto \text{pk}(sk_b)\} \\ \hline \text{new } n; \\ \text{out}\left(c, \{n\}_{\text{pk}(sk_b)}\right); \\ \text{secret}(n) \end{array} \right] \end{array} \xrightarrow{\text{out}(c, y)} \begin{array}{c} \text{new } sk_b, n; \\ \left[\begin{array}{c} \left\{ \begin{array}{l} x \mapsto \text{pk}(sk_b), \\ y \mapsto \{n\}_{\text{pk}(sk_b)} \end{array} \right\} \\ \hline \text{secret}(n) \end{array} \right] \end{array} \quad \text{(TRANSITION 4)}$$

In the state after the transition above, it is impossible to enable the secret action. That is, it is impossible to write down a message M on the label that does not use fresh variables sk_b and n (as enforced by the ALIAS rule) such that $\mathsf{M} \{x \mapsto \text{pk}(sk_b), y \mapsto \{n\}_{\text{pk}(sk_b)}\} =_{\mathsf{E}} m$ holds. This means that the message n is indeed secret for this example, i.e., secrecy is satisfied. Although there are infinitely many choices of M for an informal argument it is sufficient to notice that n is only available to the attacker in the form of the encrypted message x, but it is impossible for the attacker to obtain the secret key. In later sections, we will provide proof calculi for checking systematically such claims.

Let statements. Let statements are sometimes used as a convenience, to define variables that are abbreviations for larger messages. The operational rule, named LET in Fig. 3.6, simply instantiates the variable with the message that the variable maps to. When applying that rule make sure that you apply the substitution everywhere in scope so that the variable used as an abbreviation no longer appears in the expression. The keyword let ... in was first popularised in ML (Meta Language) [89], which has influenced many programming languages. In some languages, e.g. Haskell [73], you may have seen the keyword where serving a similar purpose.

3.3.4 A formulation of secrecy and traces

We can precisely define what it means for a protocol not to reveal secrets.

Definition 3.2 (secrecy) A state State_A *reveals no secrets*, whenever:

- there is no M and State_B such that $\text{State}_A \xrightarrow{\text{secret}(M)} \text{State}_B$
- and, furthermore, if $\text{State}_A \xrightarrow{\pi} \text{State}_C$ then State_C reveals no secrets.

We say that a process P (e.g., a thread) reveals no secrets when secrecy holds for $[id \mid P]$, where *id* is the identity substitution.

> *Secrecy.* Consider how Def. 3.2 reflects our informal understanding of secrecy. A message term is secret whenever, once it has been established as a secret, at no point in any execution of the protocol will an attacker have enough information to access that message term. The SECRET rule in Fig. 3.6 captures when an attacker can compromise a secret, as the terms appearing on the label represent what the attacker has access to. If they will never have access to the secret, such a transition will never occur.

When there is an attack on secrecy, it is relatively easy to describe the attack as a *trace*, which is a sequence of actions. A trace is an attack on secrecy is one which contains a secret action. We can represent traces using a modal logic formula using the following grammar, where π ranges over all action label as defined in SYNTAX OF LABELS.

$\phi ::= \langle \pi \rangle \phi$ It is possible to perform action π to reach a state satisfying ϕ.
$\quad\mid \text{true}$ The logical constant true.

(SYNTAX OF TRACES)

The diamond modality $\langle \pi \rangle \phi$ is used to say that it is possible to perform an action, where π is a label of the labelled transition system, i.e., $\text{in}(M, N)$, $\text{out}(M, x)$ or

3.3 Modelling protocol behaviour using operational semantics

secret(M). The logical constant true always holds, whether or not there are further actions to perform in the state that is reached by that point.

Definition 3.3 For a state State_A and a formula ϕ, satisfiability $\mathsf{State}_A \models \phi$ is defined as follows:

- $\mathsf{State}_A \models \langle \pi \rangle \phi$ holds whenever there exists State_B such that $\mathsf{State}_A \xrightarrow{\pi} \mathsf{State}_B$ and $\mathsf{State}_B \models \phi$.
- $\mathsf{State}_A \models \mathsf{true}$ holds.

Whenever $\mathsf{State}_A \models \phi$ holds we say that state State_A satisfies formula ϕ.

This leads us to the following theorem which provides us with an alternative definition of secrecy.

Theorem 3.1 (secrecy characterisation) *A state State_A can reveal a secret (i.e., it is not the case that a state reveals no secret) whenever for some $\pi_1, \ldots \pi_n$, and* M *we have* $\mathsf{State}_A \models \langle \pi_1 \rangle \cdots \langle \pi_n \rangle \langle secret(\mathsf{M}) \rangle \mathsf{true}$.

3.3.4.1 An example attack expressed as a trace

We now put together the examples presented so far to show how attacks can be expressed precisely using the given machinery. The attack in our running example in this section can be described using the formula on the right of the satisfaction relation below.

$$\mathsf{new}\ sk_a; \begin{bmatrix} \dfrac{\{x \mapsto \mathsf{pk}(sk_a)\}}{\mathsf{new}\ m;} \\ \mathsf{out}\bigl(c, \{m\}_{\mathsf{pk}(sk_e)}\bigr); \\ \mathsf{in}(c, z); \\ \mathsf{if}\ \mathsf{dec}(z, sk_a) = m\ \mathsf{then} \\ \mathsf{secret}(m) \end{bmatrix} \models \begin{array}{l} \langle \mathsf{out}(c, y) \rangle \\ \langle \mathsf{in}(c, \{\mathsf{dec}(y, sk_e)\}_x) \rangle \\ \langle \mathsf{secret}(\mathsf{dec}(y, sk_e)) \rangle \\ \mathsf{true} \end{array}$$

The above satisfaction holds, meaning that, starting in the given state, we can perform the actions $\mathsf{out}(c, y)$, and then $\mathsf{in}(c, \{\mathsf{dec}(y, sk_e)\}_x)$ in order to reach a state in which a secret is revealed, as witnessed by the event $\mathsf{secret}(\mathsf{dec}(y, sk_e))$.

You can see in the example above how the notation adopted cleanly separates between the representation of honest agents and the perspective of the attacker. The internal state of honest participants and the messages they have sent appears as an extended process on the left. The actions from the perspective of the attacker appear on the right. This clean separation between the perspectives of honest agents and attackers helps with preventing infeasible attacker capabilities to be written down, such as reversing one-way hash functions, or using private keys unavailable to the attacker. This is a hallmark of the methodology encouraged by this book, and is grounded in decades of research in concurrency theory, where the semantics of processes is defined in terms of what is observable from the environment interacting with the process rather than what happens internally [90]. This maps neatly to the security setting of the applied π-calculus, where the observer is the attacker on the network, and the semantics of processes is given in terms of potential attack vectors comprising input and output actions that the attacker can perform without looking at the internal state.

3.3.4.2 An example proof

For readers familiar with inductive definitions in mathematics, e.g., for proving properties about natural numbers, notice there is no base case in Definition 3.2. It is not an inductive proof; in fact we are employing here a technique called co-induction. In order to prove formally the above property, you should build a set of states, that in general will be infinite, such that the following holds:

- firstly, the set of states must contain the initial state, that is, the process that we are checking;
- secondly, is the set of states is preserved under all transitions in the sense that if State_A is in the set of states and $\text{State}_A \xrightarrow{\pi} \text{State}_B$, then State_B is in the set of states; and
- no state in the set reveals a secret (this is a co-inductive invariant).

We revisit the example extended processes presented when discussing TRANSITION 4, which is relatively simple compared to realistic examples. To prove that no secret is leaked, it is sufficient to note that the following set consisting of two states is indeed closed under all transitions, and, as argued previously, neither state can perform a secret transition.

$$\left\{ \begin{array}{l} \text{new } sk_b; \\ \left[\dfrac{\{x \mapsto \text{pk}(sk_b)\}}{\begin{array}{l}\text{new } n; \\ \text{out}\left(c, \{n\}_{\text{pk}(sk_b)}\right); \\ \text{secret}(n)\end{array}} \right] \end{array} , \begin{array}{l} \text{new } sk_b, n; \\ \left[\dfrac{\left\{\begin{array}{l} x \mapsto \text{pk}(sk_b), \\ y \mapsto \{n\}_{\text{pk}(sk_b)} \end{array}\right\}}{\text{secret}(n)} \right] \end{array} \right\}$$

However, for any interesting example (the above is too simple and is just provided as a minimal example of a secrecy proof), there are many sources of infinity: notably there are infinitely many inputs to choose from for an input transition, and we will move to the setting where we have infinitely many threads in the next section. Tools such as Scyther [47], Proverif [28], DEEPSEC [41] and Tamarin [87] make use of various techniques for dealing with such sources of infinity.

3.3.5 Practice in operational semantics for threads

Previously, we claimed that it is trivial that the secrecy of message m is not preserved by the responder in the Dolev-Yao 1 protocol in Fig. 3.4, under any reasonable

3.3 Modelling protocol behaviour using operational semantics

assumptions. Prove this claim by showing that, if we include a secrecy event at the end of the protocol, the attacker can always trigger the secrecy event demonstrating that m has been revealed, even if responses are only served to an honest agents. That is, the responder is only ready to talk to a specific initiator (which models a simpler threat model than we will encounter in future chapters). The names of the agents a and b running the initiator and responder roles respectively are free variables, which makes them known to the attacker. This problem statement is made precise in the following exercise.

Exercise 3.3 Show three labelled transitions beginning from the following state. Select carefully the input at the first step, representing the choice of input of the attacker, so that the final secret revealing action may be performed.

$$\left[\text{new } sk_a, sk_b; \left\{ \begin{array}{l} y \mapsto \text{pk}(sk_a), \\ z \mapsto \text{pk}(sk_b) \end{array} \right\} \left| \begin{array}{l} \text{in}(c,x); \\ \text{if } \texttt{fst}(x) = a \text{ then} \\ \text{let } m = \texttt{dec}(\texttt{snd}(x), sk_b) \text{ in} \\ \text{out}\left(c, \left(b, \{m\}_{\text{pk}(sk_a)}\right)\right); \\ \text{secret}(m) \end{array} \right. \right]$$

Show all steps in the derivation of rules to improve familiarity with reading the rules of the operational semantics, which are used throughout this book. Write down a modal logic formula describing your attack.

You can check your formula above by modelling the process in ProVerif and comparing the sequences of message in the attack trace discovered by ProVerif to your modal logic formula.

type seckey.
type pubkey.
fun pk(seckey): pubkey.
fun aenc(bitstring, pubkey): bitstring.
reduc forall x: bitstring, y: seckey; adec(aenc(x, pk(y)), y) = x.

event confidential(bitstring). *(* secrecy query *)*
query m: bitstring;
  event(confidential(m)) ∧ *attacker*(m).

free c: channel.
free a, b: bitstring.
process
  new *ska*: seckey;
  new *skb*: seckey;
  out(c, pk(*ska*)); *(* first advertise keys *)*
  out(c, pk(*skb*));
  in(c, (*x1*: bitstring, *x2*: bitstring));
  let *m*: bitstring = adec(*x2*, *skb*) **in**
  out(c, (b, aenc(*m*, pk(*ska*))));

event confidential(m)

The syntax is a little different, since ProVerif avoids using the built-in projections such as fst(.) and snd(.) directly. Instead, pairs are decomposed by receiving a tuple of messages or using a let statement where the left side of the equality is a tuple. Despite small differences, the trace you observe should be very similar to the modal logic formula that you obtain. Confirm this. This code may be useful to adapt to confirm your understanding of other examples.

Exercise 3.4 (Extra practice) To conclude, we present some additional threads for practice when attempting to understand or revise the operational rules of the applied π-calculus, as defined in this section.

1. The following thread helps understand the interplay between keys known to the attacker and those hidden from the attacker.

$$\text{new } m, sk_a; \text{out}\left(c, \{m\}_{\text{pk}(sk_a)}\right); \text{out}\left(c, \{sk_a\}_{\text{pk}(sk_e)}\right); \text{secret}(m)$$

The above can perform two output transitions and a secret transition. Can you present the derivation in full of all three transitions? Hint: if $\text{out}(c, u_1)$ and $\text{out}(c, u_2)$ are the labels of the two output transitions, then the secret action should be of the form $\text{secret}(\text{dec}(u_1, \text{dec}(u_2, sk_e)))$.

2. Consider the following thread.

$$\text{new } m, sk_a; \text{out}\left(c, \{\{(hello, m)\}_{\text{pk}(sk_a)}\}_{\text{pk}(sk_e)}\right);$$
$$\text{in}(c, x); \text{if } \text{fst}(\text{dec}(x, sk_a)) = hello \text{ then } \text{out}(c, \text{snd}(\text{dec}(x, sk_a))); \text{secret}(m)$$

The above thread can perform a sequence of four transitions: an output, an input, an output and a secret. Can you provide in full the derivation tree for all four transitions. *Hint:* We must be careful to chose the input label such that the final two transitions can be performed. If the first transition is labelled $\text{out}(c, u_1)$ the input label should be $\text{in}(c, \text{dec}(u_1, sk_e))$. Notice, while you write these derivation, that this input label uses only free variable c, u_1 and sk_e, although, in the premise of the ALIAS rule, it becomes $\text{in}\left(c, \text{dec}\left(\{\{(hello, m)\}_{\text{pk}(sk_a)} \text{ pk}(sk_e)\}, sk_e\right)\right)$ due to applying the active substitution to the label.

3. Suppose we modify the previous thread as follows, where the first output occurs later.

$$\text{new } m, sk_a; \text{in}(c, x); \text{if } \text{fst}(\text{dec}(x, sk_a)) = hello \text{ then } \text{out}(c, \text{snd}(\text{dec}(x, sk_a)));$$
$$\text{out}\left(c, \{\{(hello, m)\}_{\text{pk}(sk_a)}\}_{\text{pk}(sk_e)}\right); \text{secret}(m)$$

Explain why the above thread is blocked after performing the first input transition, regardless of the input label.

3.4 From threads to networks

We now extend our syntax of processes from Sec. 3.2.1 so that we may express networks of threads, not only a single agent. We also introduce an operational semantics for networks in which threads can execute concurrently. This will be essential for defining realistic threat models in subsequent chapters, where many threats may run in parallel. Networks, can consist of many threads running in parallel and furthermore allow threads to be executed infinitely often. Therefore, we extend the syntax of processes with a notation for parallel composition as follows.

P, Q ::= 0	end	
| secret(M)	secret	
| out(M, N);P	send	
| in(M, y);P	receive	
| if $M = N$ then P	match	(SYNTAX OF NETWORKS)
| if $M = N$ then P else Q	conditional	
| new x; P	fresh	
| let x = M in T	let	
| P | Q	parallel	
| !P	replication	

In the above syntax for networks, not only to have threads, as defined previously, but those threads may be composed in parallel using terms of the form P | Q, representing that P and Q are executing concurrently. When processes execute concurrently, the actions of each process may happen in any order with respect to each other, but the order given by each of the threads individually is respected. We also have a construct for *replication* !P that means that the thread P can create infinitely many parallel copies of itself, allowing the thread to be run infinitely many times. We still have the fresh variable binder, since we still need to manage the hiding of private information such as keys and nonces appearing in networks.

The grammar for networks is *ambiguous*, meaning that one concrete term may be parsed in multiple ways, unless we provide brackets. For some terms it does not matter, for example it does not matter whether P | Q | R is interpreted as (P | Q) | R or P | (Q | R) – either way the meaning is that we have three sub-networks P, Q, and R that run concurrently with each other. To resolve ambiguities that matter, we define an operator precedence, in terms of whether an operator binds more tightly than another, i.e., we write the brackets around one operator before the other. ProVerif ensures "P | Q" binds mode tightly than "new x; P". For example, "new x;P | Q" means "new x;$(P \mid Q)$".

The definition of free variables extends to networks as follows.

fv(P | Q) = fv(P) ∪ fv(Q)   fv(!P) = fv(P)   (FREE VARIABLES OF NETWORKS)

Capture-avoiding substitution and α-conversion are extended to networks in the obvious way, such that the only binder is fresh. Definitions are provided here.

$(P \mid Q)\theta = P\theta \mid Q\theta$ $(!P)\theta = !(P\theta)$ (SUBSTITUTION FOR NETWORKS)

Notions such as α-conversion are as in Sec. 3.2.2.3 with respect to these extended definitions.

3.4.1 An operational semantics for networks

Rules for extending our labelled transition system to networks are presented in Fig. 3.8. It is important to emphasise that what we model directly represents an honest portion of the network, i.e., threads that are run by honest agents that are trusted to play honestly by the rule book. The rest of the network including the attacker is modelled by the interactions that appear on the labels, which remain exactly the same as for threads: there are inputs, outputs and an assertion for revealing secrets.

$$\frac{P \xrightarrow{\pi} \text{new } z; [\sigma \mid R] \qquad \text{dom}(\sigma) \cup z \,\#\, Q}{P \mid Q \xrightarrow{\pi} \text{new } z; [\sigma \mid R \mid Q]} \text{ (Par-l)}$$

$$\frac{Q \xrightarrow{\pi} \text{new } z; [\sigma \mid R] \qquad \text{dom}(\sigma) \cup z \,\#\, P}{P \mid Q \xrightarrow{\pi} \text{new } z; [\sigma \mid P \mid R]} \text{ (Par-r)}$$

$$\frac{P \xrightarrow{\pi} \text{new } z; [\sigma \mid R] \qquad \text{dom}(\sigma) \cup z \,\#\, P}{!P \xrightarrow{\pi} \text{new } z; [\sigma \mid R \mid !P]} \text{ (Rep-act)}$$

Fig. 3.8 Extending the labelled transition system rules of Fig. 3.6 to networks using the Par-l, Par-r and Rep-act rules.

The other rules such as Extrude and Alias for managing fresh names do not change at all compared to Fig. 3.6. The new rules are for parallel composition and replication only, which we describe next.

3.4.1.1 The Par rules

There are two Par rules, each of which allows the process either on the left or the right of a parallel composition to act. The actions are determined by the threads that appear in the process; hence the rule has a premise which looks deeper into the network to pick out a thread that is performing the action. This way the order

3.4 From threads to networks

of actions in a thread will be preserved, but the actions of two parallel threads may happen in any order with respect to each other.

For an example of the PAR rules in action, assume we have the following extended process containing a network with two parallel components.

$$\text{State}_A \triangleq \text{new } sk_b; \left[\frac{\{pk_b \mapsto \text{pk}(sk_b)\}}{\text{new } m; (\text{out}(c,m) \mid \text{in}(c,x); \text{if } \text{dec}(x,sk_b) = m \text{ then } \text{secret}(m))} \right]$$

We will explore next three transitions that can be performed starting in this state. The first of these is the output transition displayed below.

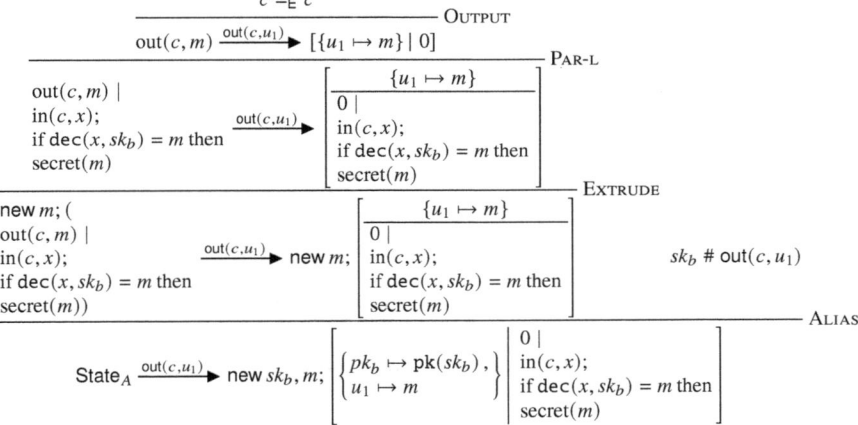

The next transition we consider starts in the state on right of the transition above, which we will call State_B. We can then perform the following input transition in state State_B, this time using PAR-R.

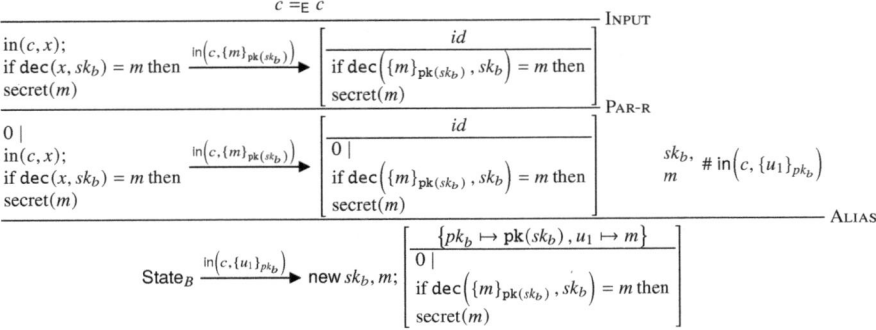

From the state we reach above, a secret transition is enabled, also by using the PAR-R rule. We make some observations about the above example. Firstly, if you try to perform the input before the output, then it is impossible that a secret transition will be enabled. This is because the input uses the output in an essential way: the output named u_1 is encrypted. Thus it can be important in which order we fire parallel components (the possible choices are said to be the *interleavings* of the process). Secondly, in the above we observe that decrypting something doesn't mean that it was always encrypted, since the attacker added the encryption layer.

The premise "$\text{dom}(\sigma)$, $z \# P$" used in the P AR rules does two things. The condition "$\text{dom}(\sigma) \# P$" ensures that the fresh name created as an alias for an output is also fresh for the parallel components of the network. The condition "$z \# P$" ensures that names extruded by one process, indicated by "new $z;\ldots$" do not accidentally capture names in the other process, thereby forcing us to carefully chose the names of extruded variables when we apply the E XTRUDE rule. For example, to enable the transition below, we were forced to rename the bound name y to something fresh, say z, which we achieve here using the E XTRUDE rule.

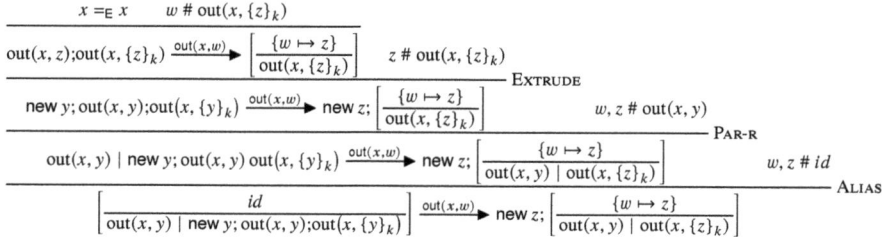

Observe above that the conditions on P AR -L, O UTPUT and A LIAS together ensure that alias w is chosen fresh everywhere. The above example transitions show two parallel threads, where the second thread only performs a transition.

3.4.1.2 The R EP - ACT rule

The idea behind replication is that we can run a replicated thread as many times as we like, and furthermore those threads may run in parallel with each other. This models the idea that there can be multiple simultaneous connections to a service offered by the thread.

Consider the following example of a state featuring a replicated thread.

$$\text{State}_C \triangleq \text{new } sk_a; \left[\begin{array}{l} \{pk_a \mapsto \text{pk}(sk_a)\} \\ \hline !\text{new } m; \\ \text{in}(c, x); \\ \text{if } \text{fst}(\text{dec}(x, sk_a)) = hi \text{ then} \\ \text{out}\left(c, \left(\text{snd}(\text{dec}(x, sk_a)), \{(hi, m)\}_{\text{pk}(sk_a)}\right)\right); \\ \text{secret}(m) \end{array} \right]$$

The above extended process can perform the input transition in Fig. 3.9. Notice the input is constructed in such a way that it would pass the next equality test in the match statement of the parallel copy of the thread created.

3.4 From threads to networks

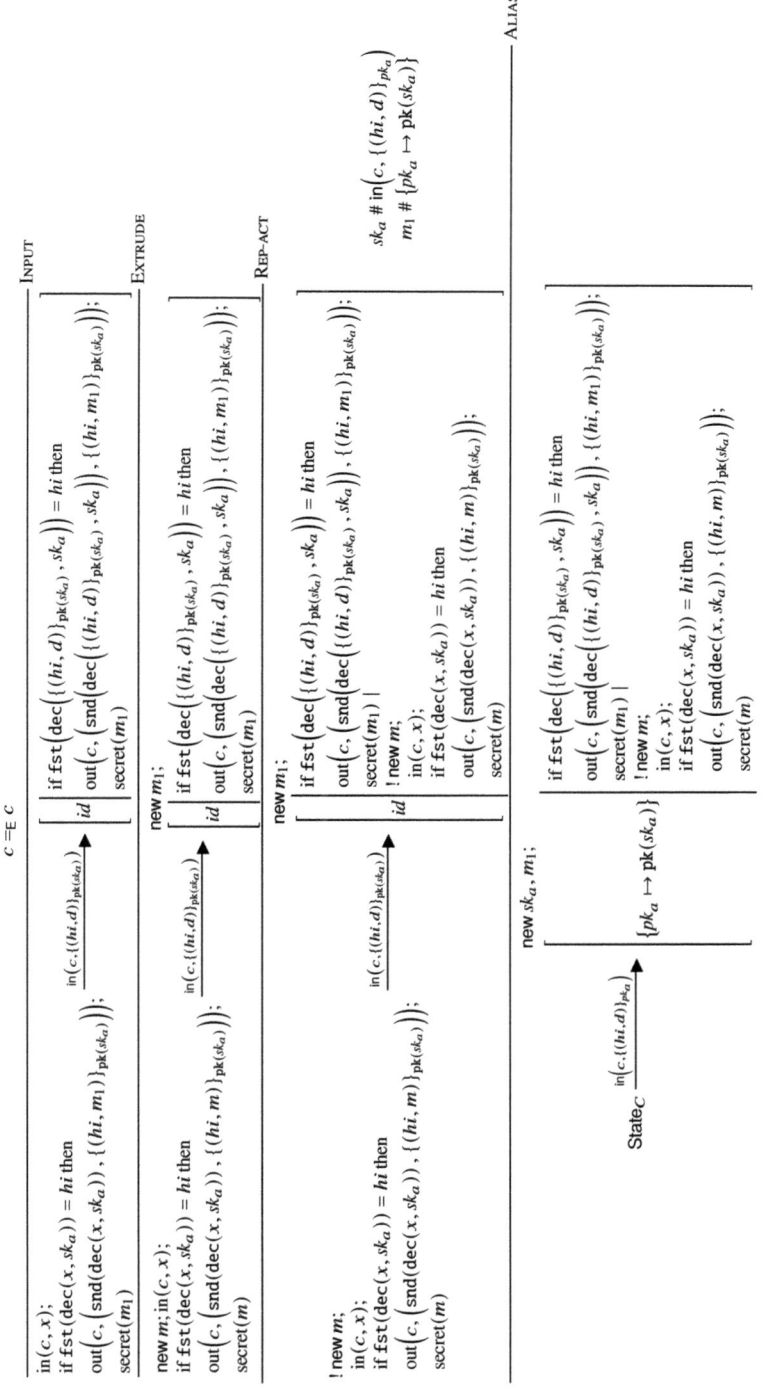

Fig. 3.9 An example involving REP-ACT.

3.4.2 Practice with the operational semantics for networks

Exercise 3.5 This advanced exercise can be used to practice how to apply the rules for networks. Consider the state that we reach by the example transition presented in Fig. 3.9, reproduced below.

$$\text{new } sk_a, m_1;$$

$$\left[\begin{array}{l} \underline{\{pk_a \mapsto \mathsf{pk}(sk_a)\}} \\ \text{if } \mathsf{fst}\Big(\mathsf{dec}\Big(\{(hi,d)\}_{\mathsf{pk}(sk_a)}, sk_a\Big)\Big) = hi \text{ then} \\ \mathsf{out}\Big(c, \Big(\mathsf{snd}\Big(\mathsf{dec}\Big(\{(hi,d)\}_{\mathsf{pk}(sk_a)}, sk_a\Big)\Big), \{(hi,m_1)\}_{\mathsf{pk}(sk_a)}\Big)\Big); \\ \mathsf{secret}(m_1) \mid \\ \quad !\,\text{new } m; \\ \quad \mathsf{in}(c,x); \\ \quad \text{if } \mathsf{fst}(\mathsf{dec}(x, sk_a)) = hi \text{ then} \\ \quad \mathsf{out}\Big(c, \Big(\mathsf{snd}(\mathsf{dec}(x, sk_a)), \{(hi,m)\}_{\mathsf{pk}(sk_a)}\Big)\Big); \\ \quad \mathsf{secret}(m) \end{array} \right]$$

The above state can perform a seqence of transitions consisting of an output action, an input action, an output action, and a secret transition.

1. Identify the first output transition and provide its derivation tree in full.
2. Consider now the state reached after the first output transition. Not all subsequent choices of input transition will lead to the secret transition being enabled. Provide the full derivation for an input transition, clearly identifying clearly the "attackers view" on the label and "the honest agent's view" represented by the extended process. When doing so, ensure that you write down the message on the input transition label that will allow the secret transition to be enabled (after a further output transition). Recall that attackers cannot see directly private names bound by fresh name binders, but can refer to free variables and variables in the domain of the substitution.
3. Write down the subsequent output and secret transitions.

Hints: The first output uses PAR-L since it involves the thread that has already started running. The second transition is an input in new session performed by using the PAR-R rule to reach the replicated thread and REP-ACT rule to create a copy of the process. The input label is input based on the previous output. The third output involves the second thread, created, so requires both PAR rules to access it. The final transition is a secret transition, but for the first copy of the thread using a recipe referring to the output from the second copy of the thread.

Chapter 4
Threat models for open networks

Abstract This chapter uses the applied π-calculus to model threats. The need to explicitly model sessions with both honest and compromised agents is justified via famous examples. Attacks are explained formally and the use of tools for automation is covered. To address the problem that adversaries can inject infinitely many inputs, the sequent calculus is introduced to calculate meaningful inputs. The operational semantics of the applied π-calculus is enhanced with constraints, defined using sequents and inequalities. The enhanced semantics is applied to calculate an attack on the secrecy of a nontrivial protocol. Cut elimination for the sequent calculus is introduced and its role in proving secrecy is illustrated on an example. Exercises apply the threat models and methods to further protocols.

4.1 Introduction

In order to model a realistic execution environment, we should reflect accurately the behaviours of honest agents and the capabilities of the attacker. The attacker, we know already, can intercept messages sent on the network and insert input messages to feed in and attempt to trick honest agents. The honest agents however we have not fully captured. We do not simply have a single honest thread that runs a finite number of actions and then stops. What really happens is that all the honest agents participating in the protocol execute concurrently in the various roles that the protocol allows (e.g., initiator and responder) and furthermore they do this as many times as they like in any combination. Beyond the fact that several agents run different roles of the protocol infinitely often there are other considerations, which we will explain and illustrate in detail in this section.

The enterprise architecture assumption. You may decide in your threat model that honest agents only communicate with agents they know are honest. This is the *outdated* enterprise security assumption where we exchange information from business to business, where there is established trust assumptions and regulatory norms that mean one business expect another businesses to adhere to. If one business is discovered to fail to meet the obligations they are likely to lose business and incur legal costs. Hence all agents trust all other agents that they communicate with, but do not trust the network in between [96].

This view is even considered to be outdated for enterprises, due to the plethora of cyber threats organisations face that could lead to a network being infiltrated, not limited to insider threats, Trojan horses, and social engineering. Hence, even enterprises when communicating with each other should have assurances that, if some of the agents in an enterprise network are compromised, then the whole organisation and their communicating partners should not be compromised by the attacker exploiting their position within the network.

The open network assumption. Assuming all communication partners are honest and uncompromised, would certainly not be appropriate for use in open networks such as the internet; or even your local network, where you cannot be certain all connections are with honest agents that are not compromised. On the internet, anyone can connect to a service and there is no guarantee that parties trust each other. Indeed, some parties connecting to a server, hereafter called *agents*, may play by the rules whereas others may not. Since in our model of security protocols we will only ever model the honest agents in the protocol, we can model this scenario by allowing threads of a protocol with honest agents to run next to open session with anyone else. Which will enable us to verify stronger security claims to the effect that if a property holds between two honest agents, then it holds even if the honest agents are engaging concurrently in many other sessions with other agents, where some of them may be attackers.

4.2 Why the enterprise architecture assumption is outdated

We consider first the outdated enterprise architecture assumption which prescribes that communication occurs only between highly regulated mutually trusting entities, such as between two banks. Indeed, in the early days of security protocols the main use case for such cryptographic protocols were within the financial sector and military. The purpose of this section is to see on a concrete example why such a threat model is not fit for purpose, so that we may better appreciate why we then go on to adapt this into a more involved threat model.

We can model such an enterprise architecture setting using the tools we have developed. In the following a and b are free variables representing the identities of two honest agents (Alice and Bob respectively). The variable c represents a channel which is used for all inputs and outputs in this model. For convenience, a separate channel *keys* for publishing public keys.

4.2 Why the enterprise architecture assumption is outdated

$\text{new } sk_a, sk_b;$
$\text{out}(keys, \text{pk}(sk_a));$ Publish the public key of a.
$\text{out}(keys, \text{pk}(sk_b)); \Big($ Publish the public key of b.
$(!Initiator(c, a, sk_a, b, \text{pk}(sk_b))) \mid$ Agent a initiates sessions with agent b.
$(!Responder(c, b, sk_b, a, \text{pk}(sk_a))) \Big)$ Agent b responds to sessions with agent a.

(ENTERPRISE 1)

You can assume that the two threads above are the initiator and responder in the Dolev-Yao 1 protocol in Fig. 3.4, as we defined previously in Sec. 3.2.4. Recall that, where it was previously defined, the initiator was parameterised as $Initiator(c, i, sk_i, r, pk_r)$. In that parametrisation, i and r were the generic roles that may be assumed, representing "initiator" and "responder" respectively, which in this case are assumed by the agents named a and b respectively. The parameter sk_i was used to represent the private key of role i, which in this case is instantiated with sk_a, which we can see is bound by a fresh name binder. What may be less immediately obvious, is that the variable in the scheme pk_r is replaced by the term $\text{pk}(sk_b)$ in the above, where sk_b, representing the private key of b is bound by a fresh name binder. This prevents the agent that assumes the initiator role from referring directly to the private key of the agent that assumes the responder role, while at the same time ensuring that the public key used by the initiator is indeed the one that corresponds to private key of the responder. The channel c plays no role in evaluating secrecy, so we keep it the same everywhere for now. A similar scheme can be applied more widely by changing initiator and responder processes for other protocols.

The first three lines of the above network simply set up the public keys, so alternatively you may prefer to start with an initial configuration where the keys are already declared. From the initial state, we can reach the following state after executing the outputs that publish the public keys.

$$\text{new } sk_a, sk_b; \left[\left\{ \begin{array}{l} pk_a \mapsto \text{pk}(sk_a), \\ pk_b \mapsto \text{pk}(sk_b) \end{array} \right\} \middle| \begin{array}{l} (!Initiator(c, a, sk_a, b, \text{pk}(sk_b))) \mid \\ (!Responder(c, b, sk_b, a, \text{pk}(sk_a))) \end{array} \right]$$

(STATE ENTERPRISE)

Now we make a claim. The Dolev-Yao 1 protocol, as modelled above, satisfies secrecy. That is, it is impossible to execute the protocol such that a secret is revealed. This fact, even for this simple protocol, is rather tricky to prove, especially using only the machinery introduced so far. So, for now, we ask the reader to either take this in faith or try to implement the problem in a tool such as ProVerif.

Code for ProVerif that verifies the above claim is provided below in four pieces, that together form a single input that ProVerif can execute. The syntax is slightly different from what you have seen and there are many type annotations; however essentially it models exactly the scenario above, as we will explain. The code can also be downloaded from Springer Nature Code Inside: https://github.com/sn-code-inside/security-protocols-and-threat-models.

ProVerif requires us to define our equational theory for messages. The following preamble defines an equational theory for public key encryption.

type pubkey.
type seckey.
fun pk(seckey): pubkey.
fun aenc(bitstring, pubkey): bitstring.
reduc forall x: bitstring, y: seckey; adec(aenc(x, pk(y)), y) = x.

The model in these notes does not require types; but we use the typed version of ProVerif when presenting code. We declare here a type for private keys (seckey) and their corresponding public keys (pubkey). The type bitstring is a generic type that can be used for ciphertext, messages, etc. There is an alternative untyped mode for ProVerif, but we consider the use of types for messages here to be intuitive enough for the reader to pick up without formal definitions for types.

In the lines below, we include some constructs that we will use to achieve the same effect as our secrecy assertion and definition in our model. This has the same effect as checking whether a secret is revealed in a trace, as in Def. 3.2.

event confidential_i(bitstring).
query m: bitstring; **event**(confidential_i(m)) ∧ *attacker*(m).
event confidential_r(bitstring).
query m: bitstring; **event**(confidential_r(m)) ∧ *attacker*(m).

The above state that, if the confidential event fires and indicates a message m should be secret, the attacker should not know m. When the attacker does know such a message, this is the same as a `secret` rule firing in our operational semantics, with the recipe for producing the secret on the label. Therefore, if this query holds then secrecy fails, and hence we want this query to be false for there to be no attack. We create two instances of the same type of event for this example, so that we can clearly distinguish between a secrecy failure on the side of the initiator or responder.

In the specifications of the roles below, we follow almost exactly the form of the roles for the Dolev-Yao 1 protocol as defined previously in Sec. 3.2.4.

let *Initiator*(c: channel, i: bitstring, *ski*: seckey, r : bitstring, *pkr*: pubkey) =
   new m : bitstring;
   out(c, (i, aenc(m, *pkr*)));
   in(c, (h : bitstring, *ms*: bitstring));
   if m = adec(*ms*, *ski*) **then**
   if $h = r$ **then**
   (* Checks if m is secret *)
   event confidential_i(m).

let *Responder*(c: channel, r : bitstring, *skr*: seckey, i: bitstring, *pki*: pubkey) =
   in(c, (h :bitstring, *ms*: bitstring));
   let m :bitstring = adec(*ms*, *skr*) **in**
   if $h = i$ **then**
   out(c, (r, aenc(m, *pki*)));

(* Checks if m is secret *)
event confidential_r(*m*).

A small stylistic difference between the code above and the syntax in this book, beyond the type annotations, is that pattern matching in inputs and **let**-statements is used to decompose tuples. Notice that we have also included another secret assertion at the end of the other thread, which has the effect of checking, secrecy of the messages received by the responder. As we know from Exercise 3.3, that thread does not preserve secrecy, so this check should confirm the trace from that exercise is an attack on secrecy.

Finally, we define the network that executes these threads below. The following defines a simplistic enterprise system where only two honest agents, *a* and *b*, communicate.

free a, b: bitstring.
free c: channel.
process
   new *ska*: seckey;
   new *skb*: seckey;
   out(c, pk(*ska*));
   out(c, pk(*skb*)); (
   (!*Initiator*(c, a, *ska*, b, pk(*skb*))) |
   (!*Responder*(c, b, *skb*, a, pk(*ska*))))

You can see above that the free variables are made explicit in ProVerif, so that their types can be indicated. The process behaves almost exactly as defined in Example ENTERPRISE 1.

If you run all the above lines together in ProVerif, you will see the result of queries, where the results indicate that the secrecy of *m* from the perspective of the initiator is preserved, but not from the perspective of the responder. We will see later that this goes against the analysis of tools with a built-in threat model such as Scyther, when we model the above protocol. The threat model employed in this section has limitations. In particular, observe that in the above network, the two honest agents may only ever talk to each other. That models a situation where there is a unique public key for each agent to talk to a specific agent. It does not, for example, cover a situation where agent *a* is using the same public key to receive messages from two different agents; which today is a normal situation – we do not generate a unique public key and distribute it to every potential communication partner.

4.3 Modelling realistic networks such as the internet

A famous example of a protocol whose security analysis breaks down with a more realistic threat model is the Needham-Schroeder protocol [95]. The Needham-

Schroeder protocol was, in fact, proven to be correct using the methods available at the time (called BAN logic), where the analysis, like our preliminary model in the subsection above, did not account for the fact that anyone could possibly connect to a service and engage in the protocol.

The researcher Gavin Lowe developed a threat model[1] and used it to more carefully account for the assumptions [81]. This was a turning point for the symbolic verification of security protocols, signalling a shift towards the process-based approach in this book. We will illustrate the modelling assumptions of Lowe, firstly, on our example Dolev-Yao 1 protocol. The analysis is slightly simpler than for the Needham-Schroeder protocol, but the essence of the problem is the same.

Consider the following alternative configuration for the Dolev-Yao 1 protocol, which features an additional inputs before each responder thread.

new sk_a, sk_b;	Declare private keys of a and b.
out($keys$, pk(sk_a));	Publish the public key of a.
out($keys$, pk(sk_b)); (	Publish the public key of b.
(!$Initiator(c, a, sk_a, b, \text{pk}(sk_b))$) |	Agent a initiates session with agent b.
(!$\underline{\text{in}(c, e)}$;	Agent b responds to anyone connecting.
if $e = a$ then	
$Responder(c, b, sk_b, a, \text{pk}(sk_a))$	If a connects, use the public key for a.
else	
$\underline{\text{in}(c, pk_e)};$	
$Responder(c, b, sk_b, e, pk_e)$))	Otherwise, use any public key.
	(OPEN ARCHITECTURE)

The first input, underlined above, receives the identity of the agent that is connecting to the responder. The conditional statement says that if the agent connecting is a known honest agent then we use their public key. Otherwise, the connecting agent is not trusted and hence could have provided anything as their public key, which is modelled above by pk_e.

> *Open networks.* We refer to the principle of modelling open sessions as the "open network" threat model. When an honest agent connects to another honest agent the correct public key are used. When an honest agent connects to anyone else, we model the worst case threat where the choice of public key is controlled fully by the attacker.

The key difference compared to the previous model is that the responder does not only talk to honest agent a; the responder is also willing to talk to anyone who provides their identity and public key.

---

[1] That original threat model of Lowe is strongly related the models in this book, but uses a slightly different style, which is based on a process calculus called CSP [101]. This book is based on the applied π-calculus [8, 70], which is also a process calculus.

4.3 Modelling realistic networks such as the internet

The if-then-else statements used in (OPEN ARCHITECTURE) above are such that they enforce a functional map from honest agent identities to public keys. This functional mapping prevents an attacker from claiming to be an honest agent, and then posting a different public key. Recall that the definition of a function as a relation is such that if $a = b$ then $f(a) = f(b)$, thus by stating the relationship between honest identities and public keys is functional, each honest identity is associated with a specific public key (or key set in some applications). This models the presence of some form of trustworthy public key infrastructure where the public key associated with honest identities cannot be reassigned by an attacker.

We make some simplifying assumptions, which work for this protocol, but not for every protocol. For this example, we can reason that the attacker gains no information by running legitimate sessions with honest initiators, since all they would learn is the secret message of the initiator. In general, for other protocols, it can matter if we also have open initiators willing to connect with anyone. When secrecy of the responder is an issue, we would also need to reflect that in the design of the configuration. The reason for this simplifying assumption here is pedagogical: we make our first example as small as possible. This threat model is sufficient to provide the following example execution of an attack in full.

An attack on the Dolev-Yao 1 protocol is expressed informally in the message sequence chart in Fig. 4.1. We will explain how to interpret this attack on the secrecy of the Dolev-Yao 1 protocol, but formally using the threat model we have developed. The agent in the middle, e, is the attacker so is implicit, i.e., it is not a thread but is instead a depiction of the actions that are used to interact with the two honest agents a and b. Therefore to interpret the above diagram we should look at the order of interactions between e and the honest agents a and b. This leads us to the attack described by the trace to the right of Fig. 4.2, which is satisfied by the state given to the left of Fig. 4.2. Notice that the given state is reachable from a state starting with the network in (OPEN ARCHITECTURE), where initially we have the identity substitution, since the first two steps simply advertise the public keys.

The trace formula in Fig. 4.2 describes the following sequence of actions:

1. The first action $\mathsf{out}(c, u_1)$ corresponds to the first message sent by agent a and intercepted by the attacker.
2. The second and third actions $\langle \mathsf{in}(c, e) \rangle$ and $\langle \mathsf{in}(c, \mathsf{pk}(sk_e)) \rangle$ correspond to the attacker connecting to the responder with a fake identity e and public key $\mathsf{pk}(sk_e)$.
3. The next input $\mathsf{in}(c, (e, \mathsf{snd}(u_1)))$ corresponds to the attacker taking the output intercepted and replacing the first component, which was a, with the fake identity e, without changing the second component.
4. There is then an output $\langle \mathsf{out}(c, u_2) \rangle$, which is the response of b to these inputs. That output is intercepted by the attacker as it travels on the network.
5. The attacker then fools a into thinking that the protocol completed successfully with b, by making use of message u_2. This is described by the input action $\langle \mathsf{in}\left(c, T\left(b, \{\mathsf{dec}(\mathsf{snd}(u_2), sk_e)\}_{pk_a}\right)\right) \rangle$, where the attacker uses their own secret key to decrypt the ciphertext (the second component of the message u_2), and

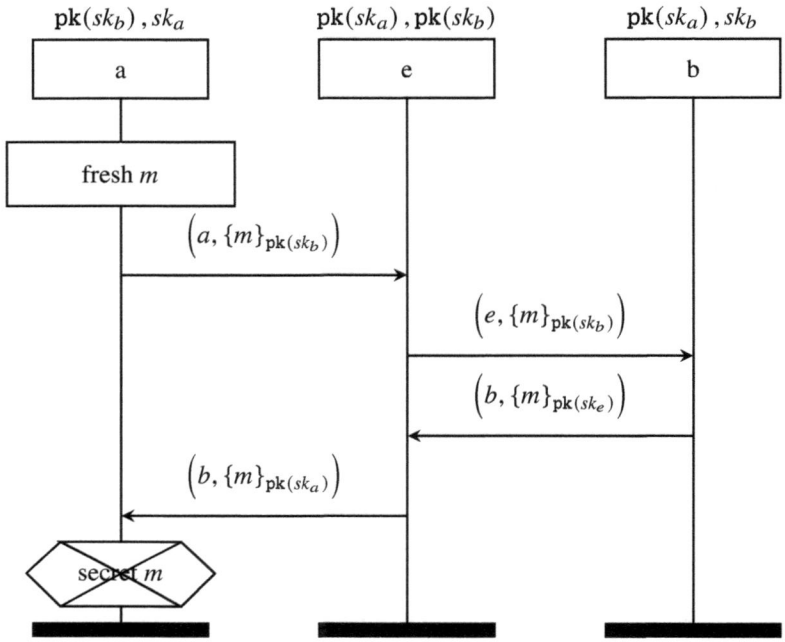

Fig. 4.1 Informal representation of attack on Dolev-Yao 1 using an MSC.

$$
\begin{bmatrix}
\text{new } sk_a, sk_b; \\
\begin{bmatrix}
\{ pk_a \mapsto \text{pk}(sk_a), pk_b \mapsto \text{pk}(sk_b) \} \\
\hline
!Initiator(c, a, sk_a, b, \text{pk}(sk_b)) \mid \\
! \text{ in}(c, e); \\
\quad \text{if } e = a \text{ then} \\
\qquad Responder(c, b, sk_b, a, \text{pk}(sk_a)) \\
\quad \text{else} \\
\qquad \text{in}(c, pk_e); \\
\qquad Responder(c, b, sk_b, e, pk_e)
\end{bmatrix}
\end{bmatrix}
\models
\begin{matrix}
\langle \text{out}(c, u_1) \rangle \\
\langle \text{in}(c, e) \rangle \\
\langle \text{in}(c, \text{pk}(sk_e)) \rangle \\
\langle \text{in}(c, (e, \text{snd}(u_1))) \rangle \\
\langle \text{out}(c, u_2) \rangle \\
\langle \text{in}(c, T(b, \{\text{dec}(\text{snd}(u_2), sk_e)\}_{pk_a})) \rangle \\
\langle \text{secret}(\text{dec}(\text{snd}(u_2), sk_e)) \rangle \\
\text{true}
\end{matrix}
$$

Fig. 4.2 The attack on Dolev-Yao 1 from Fig. 4.1, expressed formally.

4.3 Modelling realistic networks such as the internet

then encrypts it again with the public key of a, which is given by alias pk_a in the domain of the active substitution. The new ciphertext, readable by a forms the second component of the input message, while the first component is used to pretend that the message originates from b.

6. At the final step we prove that the protocol does not preserve the secrecy of m, since the message can be produced by decrypting the second message intercepted, u_2, using the fake secret key manufactured by e. This is the meaning of action $\mathrm{secret}(\mathrm{dec}(\mathrm{snd}(u_2), sk_e))$.

By following the sequence of actions described above, the Dolev-Yao 1 protocol reveals secrets for the network in (OPEN ARCHITECTURE), i.e., secrecy is violated under the "open network" threat model.

In what follows, we use the abbreviation below.

$$OpenR1(c, b, sk_b, a, pk_a) \triangleq \mathrm{in}(c, e);$$
$$\quad \text{if } e = a \text{ then}$$
$$\quad\quad Responder(c, b, sk_b, a, pk_a)$$
$$\quad \text{else}$$
$$\quad\quad \mathrm{in}(c, pk_e);$$
$$\quad\quad Responder(c, b, sk_b, e, pk_e)$$

We now provide the trace for the attack in Fig. 4.2, in full, including all transitions and all intermediate states.

$$\mathrm{new}\, sk_a, sk_b; \left[\left\{ \begin{array}{l} pk_a \mapsto \mathrm{pk}(sk_a), \\ pk_b \mapsto \mathrm{pk}(sk_b) \end{array} \right\} \middle| \begin{array}{l} !Initiator(c, a, sk_a, b, \mathrm{pk}(sk_b)) \mid \\ !OpenR1(c, b, sk_b, a, \mathrm{pk}(sk_a)) \end{array} \right]$$

$$\downarrow \mathrm{out}(c, u_1)$$

$$\mathrm{new}\, sk_a, sk_b, m_1; \left[\left\{ \begin{array}{l} pk_a \mapsto \mathrm{pk}(sk_a), \\ pk_b \mapsto \mathrm{pk}(sk_b), \\ u_1 \mapsto (a, \{m_1\}_{\mathrm{pk}(sk_b)}) \end{array} \right\} \middle| \begin{array}{l} \mathrm{in}(c, x); \\ \text{if } \mathrm{fst}(x) = b \text{ then} \\ \text{if } \mathrm{dec}(\mathrm{snd}(x), sk_a) = m_1 \text{ then} \\ \mathrm{secret}(m_1) \mid \\ (!Initiator(c, a, sk_a, b, \mathrm{pk}(sk_b))) \mid \\ (!OpenR1(c, b, sk_b, a, \mathrm{pk}(sk_a))) \end{array} \right]$$

$$\downarrow \mathrm{in}(c, e)$$

$\text{new } sk_a, sk_b, m_1; \left[\begin{Bmatrix} pk_a \mapsto \text{pk}(sk_a), \\ pk_b \mapsto \text{pk}(sk_b), \\ u_1 \mapsto \left(a, \{m_1\}_{\text{pk}(sk_b)}\right) \end{Bmatrix} \middle| \begin{array}{l} \text{in}(c, x); \\ \text{if } \text{fst}(x) = b \text{ then} \\ \quad \text{if } \text{dec}(\text{snd}(x), sk_a) = m_1 \text{ then} \\ \quad\quad \text{secret}(m_1) \mid \\ \quad\quad (!Initiator(c, a, sk_a, b, \text{pk}(sk_b))) \mid \\ \quad\quad \text{if } e = a \text{ then} \\ \quad\quad\quad Responder(c, b, sk_b, a, pk_a) \\ \quad\quad \text{else} \\ \quad\quad\quad \text{in}(c, pk_e); \\ \quad\quad\quad Responder(c, b, sk_b, e, pk_e) \mid \\ \quad\quad (!OpenR1(c, b, sk_b, a, \text{pk}(sk_a))) \end{array} \right]$

$\downarrow \text{in}(c, \text{pk}(sk_e))$

$\text{new } sk_a, sk_b, m_1; \left[\begin{Bmatrix} pk_a \mapsto \text{pk}(sk_a), \\ pk_b \mapsto \text{pk}(sk_b), \\ u_1 \mapsto \left(a, \{m_1\}_{\text{pk}(sk_b)}\right) \end{Bmatrix} \middle| \begin{array}{l} \text{in}(c, x); \\ \text{if } \text{fst}(x) = b \text{ then} \\ \quad \text{if } \text{dec}(\text{snd}(x), sk_a) = m_1 \text{ then} \\ \quad\quad \text{secret}(m_1) \mid \\ \quad\quad (!Initiator(c, a, sk_a, b, \text{pk}(sk_b))) \mid \\ \quad\quad Responder(c, b, sk_b, e, \text{pk}(sk_e)) \mid \\ \quad\quad (!OpenR1(c, b, sk_b, a, \text{pk}(sk_a))) \end{array} \right]$

$\downarrow \text{in}(c, (e, \text{snd}(u_1)))$

$\text{new } sk_a, sk_b, m_1; \left[\begin{array}{l} \begin{Bmatrix} pk_a \mapsto \text{pk}(sk_a), \\ pk_b \mapsto \text{pk}(sk_b), \\ u_1 \mapsto \left(a, \{m_1\}_{\text{pk}(sk_b)}\right) \end{Bmatrix} \\ \hline \text{in}(c, x); \\ \text{if } \text{fst}(x) = b \text{ then} \\ \quad \text{if } \text{dec}(\text{snd}(x), sk_a) = m_1 \text{ then} \\ \quad\quad \text{secret}(m_1) \mid \\ \quad\quad (!Initiator(c, a, sk_a, b, \text{pk}(sk_b))) \mid \\ \quad\quad \text{if } \text{fst}\left(\left(e, \text{snd}\left(\left(a, \{m_1\}_{\text{pk}(sk_b)}\right)\right)\right)\right) = e \text{ then} \\ \quad\quad\quad \text{out}\left(c, \left(b, \left\{\text{dec}\left(\text{snd}\left(\left(a, \{m_1\}_{\text{pk}(sk_b)}\right)\right), sk_b\right)\right\}_{\text{pk}(sk_e)}\right)\right) \mid \\ \quad\quad (!OpenR1(c, b, sk_b, a, \text{pk}(sk_a))) \end{array} \right]$

$\downarrow \text{out}(c, u_2)$

4.3 Modelling realistic networks such as the internet

$$\text{new } sk_a, sk_b, m_1; \left[\dfrac{\left\{ \begin{array}{l} pk_a \mapsto \mathsf{pk}(sk_a), \\ pk_b \mapsto \mathsf{pk}(sk_b), \\ u_1 \mapsto \left(a, \{m_1\}_{\mathsf{pk}(sk_b)}\right) \\ u_2 \mapsto \left(b, \left\{\mathsf{dec}\left(\mathsf{snd}\left(\left(e, \{m_1\}_{\mathsf{pk}(sk_b)}\right)\right), sk_b\right)\right\}_{\mathsf{pk}(sk_e)}\right) \end{array} \right\}}{\begin{array}{l} \mathsf{in}(c, x); \\ \text{if } \mathsf{fst}(x) = b \text{ then} \\ \text{if } \mathsf{dec}(\mathsf{snd}(x), sk_a) = m_1 \text{ then} \\ secret(m_1) \mid \\ (!Initiator(c, a, sk_a, b, \mathsf{pk}(sk_b))) \mid \\ 0 \mid \\ (!OpenR1(c, b, sk_b, a, \mathsf{pk}(sk_a))) \end{array}} \right]$$

$$\downarrow \mathsf{in}\left(c, \left(b, \{\mathsf{dec}(\mathsf{snd}(u_2), sk_e)\}_{pk_a}\right)\right)$$

$$\text{new } sk_a, sk_b, m_1; \left[\dfrac{\left\{ \begin{array}{l} pk_a \mapsto \mathsf{pk}(sk_a), \\ pk_b \mapsto \mathsf{pk}(sk_b), \\ u_1 \mapsto \left(a, \{m_1\}_{\mathsf{pk}(sk_b)}\right) \\ u_2 \mapsto \left(b, \left\{\mathsf{dec}\left(\mathsf{snd}\left(\left(e, \{m_1\}_{\mathsf{pk}(sk_b)}\right)\right), sk_b\right)\right\}_{\mathsf{pk}(sk_e)}\right) \end{array} \right\}}{\begin{array}{l} \text{if } \mathsf{fst}(\dagger) = b \text{ then} \\ \text{if } \mathsf{dec}(\mathsf{snd}(\dagger), sk_a) = m_1 \text{ then} \\ secret(m_1) \mid \\ (!Initiator(c, a, sk_a, b, \mathsf{pk}(sk_b))) \mid \\ 0 \mid \\ (!OpenR1(c, b, sk_b, a, \mathsf{pk}(sk_a))) \end{array}} \right]$$

$$\dagger = \left(b, \left\{\mathsf{dec}\left(\mathsf{snd}\left(\left(b, \left\{\mathsf{dec}\left(\mathsf{snd}\left(\left(e, \{m_1\}_{\mathsf{pk}(sk_b)}\right)\right), sk_b\right)\right\}_{\mathsf{pk}(sk_e)}\right)\right), sk_e\right)\right\}_{pk_a}\right)$$

$$\downarrow secret(\mathsf{dec}(\mathsf{snd}(u_2), sk_e))$$

$$\text{new } sk_a, sk_b, m_1; \left[\dfrac{\left\{ \begin{array}{l} pk_a \mapsto \mathsf{pk}(sk_a), \\ pk_b \mapsto \mathsf{pk}(sk_b), \\ u_1 \mapsto \left(a, \{m_1\}_{\mathsf{pk}(sk_b)}\right) \\ u_2 \mapsto \left(b, \left\{\mathsf{dec}\left(\mathsf{snd}\left(\left(e, \{m_1\}_{\mathsf{pk}(sk_b)}\right)\right), sk_b\right)\right\}_{\mathsf{pk}(sk_e)}\right) \end{array} \right\}}{\begin{array}{l} 0 \mid \\ (!Initiator(c, a, sk_a, b, \mathsf{pk}(sk_b))) \mid \\ 0 \mid \\ (!OpenR1(c, b, sk_b, a, \mathsf{pk}(sk_a))) \end{array}} \right]$$

A reason we do not often provide such a sequence of transitions forming attacks in full, is not just that it takes up too much space. Notice, indeed, we did not even show all the inference rules that are applied to enable transitions. The better reason is that:

generating such a transition system from a modal logic formula is computationally inexpensive; and thus the modal logic formula describing the attack can been seen as an efficiently checkable *proof certificate* for an attack. That is, given a state, calculating all the transitions with a particular label that can be performed from that state is inexpensive and hence could be independently and algorithmically verified using a formal proof assistant.

4.3.1 Other tools such as Scyther

The attack described in Fig. 4.2 can be discovered also by using other tools such as Scyther. The code for modelling the Dolev-Yao 1 protocol in Scyther is given below, along with an appropriate secrecy claim.

protocol $dy1(I, R)$ {
    role I {
        fresh m: Nonce;
        send_1$(I, R, (I, \{m\}\mathsf{pk}(R)))$;
        recv_2$(R, I, (R, \{m\}\mathsf{pk}(I)))$;
        claim_i1(I, Secret, m);
    }
    role R {
        var m: Nonce;
        recv_1$(I, R, (I, \{m\}\mathsf{pk}(R)))$;
        send_2$(R, I, (R, \{m\}\mathsf{pk}(I)))$;
    }
}

The syntax of Scyther is a little different from the syntax of processes defined previously. There are some extra features. The events are annotated with numbers just to label different send and receive events and the roles of the intended sender and receiver. Inputs are presented as a pattern to match, which makes the code similar to the message sequence chart, where the message deconstructors are implicit. Notice that the public keys are generated from the role rather than a fresh private key. In Scyther, the types of free variables are explicitly provided, i.e., "m : Nonce" makes it explicit that m is a nonce. The execution environment, i.e., a context in which we plug the roles, as illustrated in (OPEN ARCHITECTURE), is implicit, which makes Scyther an easy first tool to use.

The result of running the above code should be that the property fails to hold. The tool also provides some evidence which can be used to help reconstruct an attack on secrecy.

4.3.2 Practice in verifying secrecy in open networks

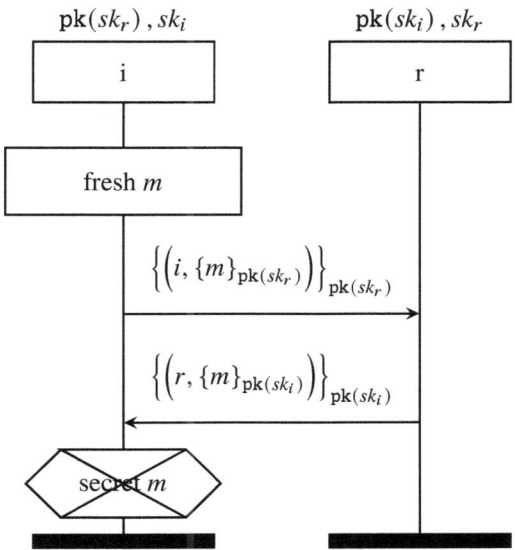

Fig. 4.3 Dolev-Yao 2: a cascade protocol, as introduced by Dolev and Yao in their seminal paper.

The original paper on the Dolev-Yao approach to symbolically verifying security protocols [48] opens with three protocols. The first we have seen and referred to as Dolev-Yao 1. The other protocols we call Dolev-Yao 2 in Fig. 4.3 and Dolev-Yao 3 in Fig. 4.4. The protocol Dolev-Yao 2 is an example of what is called a *cascade protocol*, where extra nested layers of encryption are added to the protocol using the same key. Perhaps counterintuitively, the extra layers of such cascade protocols can make them insecure, so you can break a good protocol by adding more encryption.

Exercise 4.1 To see why multiple layers of encryption can create a problem, try the following exercises. In the attack the attacker should open two sessions with an honest agent in the responder role in order to peel off the layers of encryption in the cascade protocol.

1. Model and analyse the Dolev-Yao 2 protocol, presented in Fig. 4.3, such that anyone can connect to the responder and model only initiator sessions with honest agents. To do so define the threads for each agent and provide an appropriate model of the network.

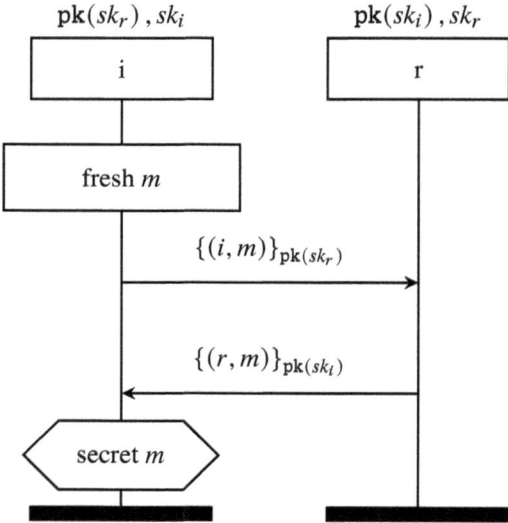

Fig. 4.4 Dolev-Yao 3: a corrected version of the protocol in Fig. 4.3, as provided by Dolev and Yao.

2. There is an attack on the Dolev-Yao 2 protocol, under this "open network" threat model. Use ProVerif to confirm this attack. You may try the same with Scyther and observe whether there is any difference between the analysis of the two tools.
3. There is however no attack on the Dolev-Yao 3 protocol, presented in Fig. 4.4. Model the Dolev-Yao 3 protocol using the tool ProVerif or Scyther to verify this claim.

In the next two sections, we will understand how the above attack and proof work, independently of the tools.

4.4 Discovering attacks with the help of proof systems

The operational semantics presented in the previous section is just right for defining when attacks exist and describing them. However, to discover an attack without knowing that it exists you can use further symbolic techniques, for calculating the inputs that result in an attack. Recall that there are infinitely many choices of input, so we cannot try them all. For this reason, when searching for an attack, it is best to treat inputs as variables that represent the most general form of inputs.

4.4.1 A sequent calculus for messages

The sequent calculus has been used since 1935 for defining the rules of various logics [61]. It is an *analytic proof system*, which basically means that it is very well behaved with respect to proof search. Indeed, for classical propositional logic, it is known that the search for a proof will always terminate. The same applies if we use the sequent calculus to calculate whether an attacker can produce a message – we will always find an answer, and, in practice, usually quickly.

A *sequent* is of the form new $\mathbf{y}$; $\Gamma \vdash$ M, where $\vdash$ is referred to as the *turnstile*. In such a sequent Γ is a comma-separated multiset of messages and M is a message that can be produced using the messages in Γ. The use of vector-like notation, $\mathbf{y}$, represents a set of variables that are local to this expression – the fresh names. We use again the keyword "new" to bind fresh names, since it is essentially the same concept as appeared in threads, networks and states. For example, the following is a sequent, with two fresh names and two messages on the left of the turnstile.

$$\text{new } sk_a, m; \text{pk}(sk_a), \left(b, \{m\}_{\text{pk}(sk_e)}\right) \vdash \left(b, \{m\}_{\text{pk}(sk_a)}\right)$$

In the context of the execution of a security protocol, the set of messages Γ represents the messages intercepted by an attacker at some point during an execution of some protocol. That is, Γ is the messages mapped to by the active substitution. The message M represents some message that should be produced, e.g., as an input or as a secret. The fresh names represent the messages that the attacker cannot access directly – the attacker can see the whole message but not necessarily all sub-terms.

$$\frac{}{\text{new } \mathbf{x}; \Gamma, \text{M} \vdash \text{M}} \text{(Ax)} \qquad \frac{z \# \mathbf{x}}{\text{new } \mathbf{x}; \Gamma \vdash z} \text{(Sol)} \qquad \frac{\text{new } \mathbf{x}; \Gamma \vdash \text{K}}{\text{new } \mathbf{x}; \Gamma \vdash \text{pk}(\text{K})} \text{(I-PK)}$$

$$\frac{\text{new } \mathbf{x}; \Gamma \vdash \text{M} \quad \text{new } \mathbf{x}; \Gamma \vdash \text{N}}{\text{new } \mathbf{x}; \Gamma \vdash (\text{M}, \text{N})} \text{(I-PAIR)} \qquad \frac{\text{new } \mathbf{x}; \Gamma \vdash \text{M} \quad \text{new } \mathbf{x}; \Gamma \vdash \text{K}}{\text{new } \mathbf{x}; \Gamma \vdash \{\text{M}\}_\text{K}} \text{(I-ENC)}$$

$$\frac{\text{new } \mathbf{x}; \Gamma, \text{M}, \text{N} \vdash \text{K}}{\text{new } \mathbf{x}; \Gamma, (\text{M}, \text{N}) \vdash \text{K}} \text{(E-PAIR)} \qquad \frac{\text{new } \mathbf{x}; \Gamma, \text{M} \vdash \text{L} \quad \text{new } \mathbf{x}; \Gamma \vdash \text{K}}{\text{new } \mathbf{x}; \Gamma, \{\text{M}\}_{\text{pk}(\text{K})} \vdash \text{L}} \text{(E-ENC)}$$

Fig. 4.5 Rules of the sequent calculus for our equational theory E.

The axioms and rules of the sequent calculus for the Dolev-Yao model are defined in Fig. 4.5, where M, N, K, L are messages. Because Γ is a multiset, formulae can be reordered on the left side of the sequent. The *premises* of a rule are the sequents above the line and the *conclusion* of the rule is the sequent below the line. A *proof* is a tree of rules, where all the leaves are axioms. For more details on how to read such rules, we refer to Section 3.3.3 where we introduced the operational semantics in the same style.

4.4.2 Examples of a deduction

The following is a proof of sequent "new $sk_b, m;$ $\left(a, \{m\}_{\text{pk}(sk_b)}\right) \vdash \left(e, \{m\}_{\text{pk}(sk_b)}\right)$" in which we adopt a convention where we pull out the fresh names to the outermost level, since they are the same in every sequent.

$$\text{new } sk_b, m; \quad \cfrac{\cfrac{\cfrac{e \;\#\; sk_b, m}{a, \{m\}_{\text{pk}(sk_b)} \vdash e} \text{Sol} \quad \cfrac{}{a, \{m\}_{\text{pk}(sk_b)} \vdash \{m\}_{\text{pk}(sk_b)}} \text{Ax}}{a, \{m\}_{\text{pk}(sk_b)} \vdash \left(e, \{m\}_{\text{pk}(sk_b)}\right)} \text{I-PAIR}}{\left(a, \{m\}_{\text{pk}(sk_b)}\right) \vdash \left(e, \{m\}_{\text{pk}(sk_b)}\right)} \text{E-PAIR}$$

Notice that the above corresponds to the first two steps of the attack in Dolev-Yao 1 protocol as shown in the diagram in Fig. 4.2. The attacker learns the message $\left(a, \{m\}_{\text{pk}(sk_b)}\right)$ and uses it to produce the message $\left(e, \{m\}_{\text{pk}(sk_b)}\right)$, without ever using directly the private variables sk_b and m, hidden by the fresh name binder. So what is happening in the proof tree above is that the sequent calculus is performing *algorithmically* the thought process that we went through previously. This process we refer to as *deduction*.

We can apply the same ideas as above to the other messages of the attack in the diagram in Fig. 4.2, in order to *deduce* the inputs from the outputs. Observe that the attacker intercepts the message $\left(b, \{m\}_{\text{pk}(sk_e)}\right)$ and already knew the public key $\text{pk}(sk_a)$ and that sk_a and m are private variables. We then wish to produce an input of the form $\left(b, \{m\}_{\text{pk}(sk_a)}\right)$. This leads us to the sequent we saw previously, namely "new $sk_a, m; \text{pk}(sk_a), \left(b, \{m\}_{\text{pk}(sk_e)}\right) \vdash \left(b, \{m\}_{\text{pk}(sk_a)}\right)$". A proof of this sequent is shown below.

$$\text{new } sk_a, m; \quad \cfrac{\cfrac{\cfrac{\cfrac{b \;\#\; sk_a, m}{\text{pk}(sk_a), b, m \vdash b} \text{Sol} \quad \cfrac{\cfrac{}{\text{pk}(sk_a), b, m \vdash m} \text{Ax} \quad \cfrac{}{\text{pk}(sk_a), b, m \vdash \text{pk}(sk_a)} \text{Ax}}{\text{pk}(sk_a), b, m \vdash \{m\}_{\text{pk}(sk_a)}} \text{I-ENC}}{\text{pk}(sk_a), b, m \vdash \left(b, \{m\}_{\text{pk}(sk_a)}\right)} \text{I-PAIR} \quad \cfrac{\cfrac{sk_e \;\#\; sk_a, m}{\text{pk}(sk_a), b \vdash sk_e} \text{Sol}}{} \text{E-ENC}}{\text{pk}(sk_a), b, \{m\}_{\text{pk}(sk_e)} \vdash \left(b, \{m\}_{\text{pk}(sk_a)}\right)} \text{E-PAIR}}{\text{pk}(sk_a), \left(b, \{m\}_{\text{pk}(sk_e)}\right) \vdash \left(b, \{m\}_{\text{pk}(sk_a)}\right)}$$

Rules are either *introduction rules* (I-PAIR and I-ENC) or *elimination rules* (E-PAIR and E-ENC). In this sequent calculus you can always apply all elimination rules first (breaking down messages on the left as much as possible); and then apply introduction rules. The sequent calculus enjoys the *sub-formula property*, meaning every formula appearing in the premises of a rule is a sub-formula of a formula in the conclusion of the rule. This means that the search for a proof will always terminate, as terms get broken down. There are many alternative proof systems, but even commonly used ones such as *natural deduction* do not enjoy such well-behaved proof search properties. For this reason we encourage the use of the sequent calculus for deduction problems. This observation is due to Dawson, Goré and Tiu [107].

4.4.3 Practice using the sequent calculus

Exercise 4.2 Prove or disprove the following sequents.

1. new $m, n, k; \{m\}_{\text{pk}(k)}, (k, n) \vdash (m, n)$.
2. new $k, m; \text{pk}(k), \{m\}_{\text{pk}(k)} \vdash m$.

Prove the following by induction on the structure of a proof. This is a *monotonicity* property which means that if the attacker intercepts new messages then the attacker can still produce messages it could previously produce.

Proposition 4.1 (monotonicity) *If* $\Gamma \subseteq \Delta$, *and* new $\mathbf{y}; \Gamma \vdash \mathsf{M}$, *and also* $\mathbf{z} \mathrel{\#} \Gamma, \mathsf{M}$, *then we have that* new $\mathbf{y}, \mathbf{z}; \Delta \vdash \mathsf{M}$ *holds.*

Hint: To prove a property of a proof tree by induction, firstly, consider the base cases which correspond to the axioms at the leaves of a tree, i.e., prove that the property holds for Ax and for Sol. For the inductive cases assume that the property holds for the premises above the line and use those assumptions to establish the property for the conclusion. Be careful with the freshness conditions for the inductive case for rule D-ENC.

4.5 Calculating input labels using the sequent calculus

In the previous section, we introduced a sequent calculus, which determines if a message can be deduced by the attacker given particular prior knowledge of the attacker, i.e., a set of messages and fresh names. We will use the sequent calculus to help calculate whether an input, induced by the attacker, follows from the outputs intercepted before the input is produced.

We will exploit here a precise correspondence between sequent calculus proofs and recipes that appear on labels of transitions in the operational semantics of the applied π-calculus. For example, consider again the following sequent:

$$\text{new } sk_b, m; \quad \cfrac{\cfrac{e \,\#\, sk_b, m}{a, \{m\}_{\text{pk}(sk_b)} \vdash e} \,\text{Sol} \qquad \cfrac{}{a, \{m\}_{\text{pk}(sk_b)} \vdash \{m\}_{\text{pk}(sk_b)}} \,\text{Ax}}{\cfrac{a, \{m\}_{\text{pk}(sk_b)} \vdash \left(e, \{m\}_{\text{pk}(sk_b)}\right)}{\left(a, \{m\}_{\text{pk}(sk_b)}\right) \vdash \left(e, \{m\}_{\text{pk}(sk_b)}\right)} \,\text{E-pair}} \,\text{I-pair}$$

If u is an alias for $\left(a, \{m\}_{\text{pk}(sk_b)}\right)$ then the above proof corresponds to the recipe $(e, \text{snd}(u))$. To see this, observe that the instance of elimination rule E-pair breaks down the pair with alias u into a first and second component, where only the second component $\text{snd}(u)$ is used in an axiom. The instance of the rule Sol acknowledges that e is a free variable and hence can be appealed to directly in a recipe, since it is known to the attacker. The rule I-pair then pairs the two message together to obtain a message forming recipe that does not refer to private names m and sk_b.

A minimal process in which this sequent can play a role in calculating a label is the following.

$$\text{new } m, sk_b; \text{out}\!\left(c, \left(a, \{m\}_{\text{pk}(sk_b)}\right)\right); \text{in}(c, x); \text{if } x = \left(e, \{m\}_{\text{pk}(sk_b)}\right) \text{ then out}(c, m)$$

Observe that there is first an output transition.

$$id \begin{bmatrix} \text{new } m, sk_b; \\ \text{out}\!\left(c, \left(a, \{m\}_{\text{pk}(sk_b)}\right)\right); \\ \text{in}(c, x); \\ \text{if } x = \left(e, \{m\}_{\text{pk}(sk_b)}\right) \text{ then} \\ \text{out}(c, m) \end{bmatrix} \xrightarrow{\text{out}(c,u)} \begin{bmatrix} \text{new } m, sk_b; \\ \cfrac{\left\{u \mapsto \left(a, \{m\}_{\text{pk}(sk_b)}\right)\right\}}{\text{in}(c, x);} \\ \text{if } x = \left(e, \{m\}_{\text{pk}(sk_b)}\right) \text{ then} \\ \text{out}(c, m) \end{bmatrix}$$

There is then some input, but such that x must pass the test $x = \left(e, \{m\}_{\text{pk}(sk_b)}\right)$ later on. Therefore, that attacker should use knowledge accumulated in the active substitution, i.e., $\left(a, \{m\}_{\text{pk}(sk_b)}\right)$, to produce something of the form $\left(e, \{m\}_{\text{pk}(sk_b)}\right)$, but without referring to m and sk_b directly which are not known to the attacker a priori. The recipe above corresponding to the proof of our running example of a sequent is exactly the required message for the input transition. The resulting input transition is illustrated below, which stems from the state reached on the right of the transition above.

$$\xrightarrow{\text{in}(c,(e,\text{snd}(u)))} \begin{bmatrix} \text{new } m, sk_b; \\ \cfrac{\left\{u \mapsto \left(a, \{m\}_{\text{pk}(sk_b)}\right)\right\}}{\text{if } \left(e, \{m\}_{\text{pk}(sk_b)}\right) = \left(e, \{m\}_{\text{pk}(sk_b)}\right) \text{ then}} \\ \text{out}(c, m) \end{bmatrix}$$

4.5 Calculating input labels using the sequent calculus

The sequent $\left(a, \{m\}_{\text{pk}(sk_b)}\right) \vdash \left(e, \{m\}_{\text{pk}(sk_b)}\right)$, solved previously and used to calculate the input label above, is obtained systematically as follows. We write down the fresh names accumulated in the extended process, the messages output in the active substitution on the left of the turnstile, and to the right of the turnstile the message unified with input variable x in order to enabled the guard. We will see that this approach applies generally to calculate input labels when input transitions that enable guards exist. Furthermore, the most general input labels are generated by the possible proofs of the sequents.[2]

4.5.1 An attack on the Dolev-Yao 2 cascade protocol: a fully worked example

We now illustrate the above technique for using sequents to calculate inputs on a larger example, where the inputs required to mount an attack are not immediately obvious. We will also make the accumulation of sequents more precise by introducing an enhanced labelled transition system that keeps track of a system of constraints defined by sequents and inequalities. The aim of this section is to show that complex patterns of inputs and outputs leading an to attack can be calculated using a powerful system and a few heuristics.

Convention. In this section, each step towards calculating the attack is highlighted as **Step #**. There will be 8 steps highlighted, interleaved with explanations.

Let us begin by formally defining the Dolev-Yao 2 protocol in Fig. 4.3. The initiator and responder can be defined as follows, much like the Dolev-Yao 1 protocol, but with an extra layer of encryption.

$I_{DY2}(c, i, sk_i, r, pk_r) \triangleq$
new m;
$\text{out}\left(c, \left\{\left(i, \{m\}_{pk_r}\right)\right\}_{pk_r}\right)$;
$\text{in}(c, x)$;
if $\text{fst}(\text{dec}(x, sk_i)) = r$ then
if $\text{dec}(\text{snd}(\text{dec}(x, sk_i)), sk_i) = m$ then
$\text{secret}(m)$

$R_{DY2}(c, r, sk_r, i, pk_i) \triangleq$
$\text{in}(c, x)$;
if $\text{fst}(\text{dec}(x, sk_r)) = i$ then
let $m = \text{dec}(\text{snd}(\text{dec}(x, sk_r)), sk_r)$ in
$\text{out}\left(c, \left\{\left(r, \{m\}_{pk_i}\right)\right\}_{pk_i}\right)$

The configuration that we use to analyse the Dolev-Yao 1 protocol, under an "open network" threat model, can also be applied here. It is sufficient for this protocol, that the responder is ready to connect to anyone including both honest agents and the attacker. We simply replace the initiator and responder processes with those defined above to obtain the following initial network configuration.

---

[2] In some examples there is more than one distinct proof and hence multiple input labels to consider are produced by this method. For this particular example, all proofs are equivalent, since the only possible change to our running example sequent proof is to delay the application of the elimination rule, which would not affect the recipe produced.

$$\begin{aligned}
&\text{new } sk_a, sk_b; \\
&\text{out}(keys, \text{pk}(sk_a)); \\
&\text{out}(keys, \text{pk}(sk_b));(\\
&\quad !I_{DY2}(c, a, sk_a, b, \text{pk}(sk_b)) \mid \\
&\quad !OpenR2(c, b, sk_b, a, \text{pk}(sk_a)) \,)
\end{aligned}$$

With *OpenR2*, we basically adapt *OpenR1* in the previous section so that it calls the Responder role of Dolev-Yao 2.

$$\begin{aligned}
OpenR2(c, b, sk_b, a, pk_a) &\triangleq \text{in}(c, e); \\
&\quad \text{if } e = a \text{ then} \\
&\qquad R_{DY2}(c, b, sk_b, a, pk_a) \\
&\quad \text{else} \\
&\qquad \text{in}(c, pk_e); \\
&\qquad R_{DY2}(c, b, sk_b, e, pk_e)
\end{aligned}$$

After two steps of execution, we reach the following state where the public keys of the two honest agents a and b have been published on the network, and hence are known to the attacker.

$$\text{State}^1_{DY2} \triangleq \text{new } sk_a, sk_b; \left[\left\{ \begin{array}{l} pk_a \mapsto \text{pk}(sk_a), \\ pk_a \mapsto \text{pk}(sk_b) \end{array} \right\} \,\Big|\, \begin{array}{l} !I_{DY2}(c, a, sk_a, b, \text{pk}(sk_b)) \mid \\ !OpenR2(c, b, sk_b, a, \text{pk}(sk_a)) \end{array} \right]$$

Now we should think about heuristics. Since there are infinitely many possible ways in which we can choose actions we should make a smart choice for which to try. A choice that frequently works is the following strategy, which we illustrate on the Dolev-Yao 2 protocol.

Step 1. Firstly, we must have an honest thread with a secret event, so we start an honest version of the initiator in role a, who talks to honest agent b. The initiator sends the first message, which, as always, the attacker intercepts.

$$\text{State}^1_{DY2} \xrightarrow{\text{out}(c, u_1)} \left[\begin{array}{l} \text{new } sk_a, sk_b, m; \\ \left\{ \begin{array}{l} pk_a \mapsto \text{pk}(sk_a), \\ pk_a \mapsto \text{pk}(sk_b) \\ u_1 \mapsto \left\{ \left(a, \{m\}_{\text{pk}(sk_b)}\right) \right\}_{\text{pk}(sk_b)} \end{array} \right\} \\ \text{in}(c, x); \\ \text{if } \text{fst}(\text{dec}(x, sk_a)) = b \text{ then} \\ \text{if } \text{dec}(\text{snd}(\text{dec}(x, sk_a)), sk_a) = m \text{ then} \\ \text{secret}(m) \mid \\ !I_{DY2}(c, a, sk_a, b, \text{pk}(sk_b)) \mid \\ !OpenR2(c, b, sk_b, a, \text{pk}(sk_a)) \end{array} \right]$$

In general, the heuristic we are applying is that any enabled output can be triggered. This is a safe strategy, since the message to output is already determined at this point and hence nothing in the system can influence that output.

4.5 Calculating input labels using the sequent calculus

Step 2. Secondly, there are several choices of input to perform, and perhaps one choice will expose an attack, unless the protocol is very badly broken. The attacker connects to an honest agent b and starts an honest responder session. The attacker uses a new fake identity, new key and initiates an input. At this point, we employ a new trick. For each of the three inputs we use a new fresh variable rather choosing a particular message. So, we perform three transitions. The first transition is labelled with $\text{in}(c, e)$ as shown below.

$$\downarrow \text{in}(c,e)$$

$$\text{new } sk_a, sk_b, m; \left[\left\{ \begin{array}{l} pk_a \mapsto \text{pk}(sk_a), \\ pk_a \mapsto \text{pk}(sk_b) \\ u_1 \mapsto \left\{ \left(a, \{m\}_{\text{pk}(sk_b)}\right)\right\}_{\text{pk}(sk_b)} \end{array} \right\} \left| \begin{array}{l} \text{in}(c,x); \\ \text{if } \text{fst}(\text{dec}(x, sk_a)) = b \text{ then} \\ \text{if } \text{dec}(\text{snd}(\text{dec}(x, sk_a)), sk_a) = m \text{ then} \\ \text{secret}(m) \mid \\ !I_{DY2}(c, a, sk_a, b, \text{pk}(sk_b)) \mid \\ \text{if } e = a \text{ then} \\ \quad R_{DY2}(c, b, sk_b, a, pk_a) \\ \text{else} \\ \quad \text{in}(c, pk_e); \\ \quad R_{DY2}(c, b, sk_b, e, pk_e) \\ !OpenR2(c, b, sk_b, a, \text{pk}(sk_a)) \end{array} \right. \right]$$

At this point, for the thread of the responder to proceed further we have to induce an assumption that $e \neq a$ in order to trigger the else branch of the if-then-else statement. This is a symbolic constraint on input variables that we can use in an enhanced transition system. We introduce a symbolic constraint system and explain how to use them next.

In Fig. 4.6, we present variations on our usual rules for the two rules impacting the constraint system. These rules are O-IN-ALIAS and O-MISMATCH. Other rules, enhanced with constraints are presented for convenience, although they change little. We explain the rules below, but first explain the constraint system we employ.

Definition 4.1 (constraint systems) A constraint system is of the form new $\mathbf{x}; C$, where C is a set of sequents and inequalities between messages of the form

$$\Gamma_1 \vdash M_1; \Gamma_2 \vdash M_2; \ldots; N_1 \neq L_1; N_2 \neq L_2; \ldots .$$

We use semicolons as separators, since commas are already used to separate the messages within the individual sequents.

We say a constraint system new $\mathbf{x}; C$ is solvable whenever both of the following conditions hold:

- all sequents $\Gamma \vdash M$ in C are such that new $\mathbf{x}; \Gamma \vdash M$ is provable; and
- for all inequalities $M \neq N$ in C, it is the case that $M =_E N$ does not hold.

We say that new $\mathbf{x}; C$ can access new $\mathbf{x}; \mathcal{D}$ by *idempotent* substitution θ whenever new $\mathbf{x}; \mathcal{D}$ has a solution where $C\theta \cup \{M_1 \neq N_1; M_2 \neq N_2; \ldots\} = \mathcal{D}$ and $\text{dom}(\theta) \# \mathbf{x}$ (i.e., fresh names cannot be changed by the substitution). Here $C\theta$ is defined in the obvious way by applying the substitution to all messages appearing in C. This definition of one constraint system accessing another captures possible ways that variables are permitted to be instantiated and also allows new inequalities to be assumed, such as $e \neq a$.

We make use of the judgement new $\mathbf{x}; C \models M \neq N$, defined whenever there exists no θ such that new $\mathbf{x}; C$ can access some other set of constraints by substitution θ and $M\theta =_\mathsf{E} N\theta$ holds.

The notion of accessibility defined above will also be used to model how we can progressively learn about the inputs lazily, that is, just at the point where we use the input value as opposed to at the point of input. The nontrivial use of accessibility in this way is illustrated throughout the running example in this section. Notice that the definitions above mean that an inequality between messages holds with respect to a constraint system when there is no possible way of instantiating variables such that the messages are equal, and hence the messages are most certainly not equal.

Intuitionism and constraints. For readers who are curious about logic, these definitions for a constraint system provide an intuitionistic interpretation of negation, where the law of the excluded middle $M = N$ or $M \neq N$ does not hold with respect to all constraint systems and messages [68, 10]. It is possible that neither $M =_\mathsf{E} N$ nor $M \neq N$ holds for some constraint system. For example, trivially $x =_\mathsf{E} y$ does not hold, and also new $x; x \vdash y \models x \neq y$ does not hold. To see why the latter does not hold, observe that if we apply unifying substitution $\{y \mapsto x\}$, that makes $x =_\mathsf{E} y$ hold, to the constraint $x \vdash y$, then we have to check whether new $x; x \vdash x$ is provable whether the accessible constraint has a solution. Indeed new $x; x \vdash x$ is provable, by using Ax, and hence it is possible that $x = y$ for some accessible constraint system.

In contrast to the above, observe that new $x; \vdash y \models x \neq y$ does hold, since the same unifier $\{y \mapsto x\}$ leads us to evaluating the provability of new $x; \vdash x$, which does not hold, since neither Sol nor Ax applies. There is a clear process intuition here. The knowledge to the left of the turnstile in new $x; \vdash y$ is empty, modelling that no message has been output at the moment y is input, and since, x is freshly generated it is impossible that x could be input in place of y. For another example, new $x, z; C \models x \neq z$ holds for any C, since there is no substitution θ unifying x and z such that dom(θ) # x, z. Fresh names cannot be unified.

The rule o-In-Alias in Fig. 4.6 updates a constraint system with a sequent that remembers the outputs known to the attacker at the point of input, i.e., the knowledge available to the attacker at that moment. The sequent constrains the infinitely many possible messages that the input variable represents. The rule using constraints is the o-Mismatch rule in Fig. 4.6 that evaluates the else branch of an if-then-else statement. The rule o-Mismatch varies with respect to the Mismatch rule already presented in that the inequality must hold with respect to a constraint system. Such an inequality can be induced by an extra explicit inequality constraint introduced

4.5 Calculating input labels using the sequent calculus

$$\frac{\text{new } \mathbf{x}; C : T_0 \xrightarrow{\pi} \text{State} \qquad \text{new } \mathbf{x}; C \models M \neq N}{\text{new } \mathbf{x}; C : \text{if } M = N \text{ then } T_1 \text{ else } T_0 \xrightarrow{\pi} \text{State}} \text{ (o-Mismatch)}$$

$$\frac{\text{new } \mathbf{x}; C : T_0 \xrightarrow{in(M\theta, N\theta)} \text{new } \mathbf{y}; [id \mid T_1] \qquad \mathbf{x} \# in(M, N) \qquad \mathbf{y} \# C, \theta}{\text{new } \mathbf{x}; C \,[\theta \mid T_0] \xrightarrow{in(M, N)} \text{new } \mathbf{x}, \mathbf{y}; C; \text{ran}(\theta) \vdash N\theta \,[\theta \mid T_1]} \text{ (o-In-Alias)}$$

$$\frac{\text{new } \mathbf{x}; C : T_0 \xrightarrow{out(M\theta, u)} \text{new } \mathbf{y}; [\sigma \mid T_1] \qquad \mathbf{x} \# out(M, u) \qquad u, \mathbf{y} \# C, \theta}{\text{new } \mathbf{x}; C \,[\theta \mid T_0] \xrightarrow{out(M, u)} \text{new } \mathbf{x}, \mathbf{y}; C \,[\theta \circ \sigma \mid T_1]} \text{ (o-Out-Alias)}$$

$$\frac{K =_E M \qquad u \# K, M, N, T}{\text{new } \mathbf{x}; C : \text{out}(M, N); T \xrightarrow{out(K, u)} [\{u \mapsto N\} \mid T]} \text{ (o-Output)}$$

$$\frac{K =_E M}{\text{new } \mathbf{x}; C : \text{in}(M, x); T \xrightarrow{in(K, N)} [id \mid T \{x \mapsto N\}]} \text{ (o-Input)}$$

$$\frac{\text{new } \mathbf{x}, z; C : T \{y \mapsto z\} \xrightarrow{\pi} \text{State} \qquad z \# C, \pi, \text{new } y; T}{\text{new } \mathbf{x}; C : \text{new } y; T \xrightarrow{\pi} \text{new } z; \text{State}} \text{ (o-Extrude)}$$

$$\frac{\text{new } \mathbf{x}; C : T_1 \xrightarrow{\pi} \text{State} \qquad M =_E N}{\text{new } \mathbf{x}; C : \text{if } M = N \text{ then } T_1 \text{ else } T_0 \xrightarrow{\pi} \text{State}} \text{ (o-Match)}$$

$$\frac{\text{new } \mathbf{x}; C : P \xrightarrow{\pi} \text{new } \mathbf{z}; [\sigma \mid R] \qquad \text{dom}(\sigma) \cup \mathbf{z} \# Q}{\text{new } \mathbf{x}; C : P \mid Q \xrightarrow{\pi} \text{new } \mathbf{z}; [\sigma \mid R \mid Q]} \text{ (o-Par-l)}$$

$$\frac{\text{new } \mathbf{x}; C : Q \xrightarrow{\pi} \text{new } \mathbf{z}; [\sigma \mid R] \qquad \text{dom}(\sigma) \cup \mathbf{z} \# P}{\text{new } \mathbf{x}; C : P \mid Q \xrightarrow{\pi} \text{new } \mathbf{z}; [\sigma \mid P \mid R]} \text{ (o-Par-r)}$$

$$\frac{\text{new } \mathbf{x}; C : P \xrightarrow{\pi} \text{new } \mathbf{z}; [\sigma \mid R] \qquad \text{dom}(\sigma) \cup \mathbf{z} \# P}{\text{new } \mathbf{x}; C : !P \xrightarrow{\pi} \text{new } \mathbf{z}; [\sigma \mid R \mid !P]} \text{ (o-Rep-act)}$$

$$\frac{N =_E M}{\text{new } \mathbf{x}; C : \text{secret}(M) \xrightarrow{secret(N)} [id \mid 0]} \text{ (o-Secret)}$$

Fig. 4.6 Rules of a labelled transition system annotated with constraint systems.

at any point to the constraint system, such as $e \neq a$, to ensure that these variables are never unified in any solution. All other rules are annotated with the constraint system, allowing the constraints to be appealed to when evaluating the inequalities in the O-MISMATCH rule, as demonstrated in Fig. 4.6. Notice that in some rules in Fig. 4.6 freshness is extended to constraint systems defined in the obvious way to apply to all messages in the constraint system. Explicitly we have $x \# C$ whenever, for all $\Gamma \vdash M$ in C, we have that $x \# M$ and for all N in Γ, $x \# N$ holds, and, also, for all $K \neq L$ in C, we have $x \# K, L$.

The constraint system for the initial network state is $\vdash x_1; \vdash x_2; \ldots \vdash x_n$, where $x_1, \ldots x_n$ are the free variables of the process modelling the network. This is to ensure that variables that are free before anything is output will never be instantiated with knowledge derived from anything that is output. So, for example, then we have the following initial state where C_0 is defined to be $\vdash c; \vdash a; \vdash b$.

$$\text{new } sk_a, sk_b; C_0 \left[\left\{ \begin{array}{l} pk_a \mapsto \mathsf{pk}(sk_a), \\ pk_a \mapsto \mathsf{pk}(sk_b) \end{array} \right\} \middle| \begin{array}{l} !I_{DY2}(c, a, sk_a, b, \mathsf{pk}(sk_b)) \mid \\ !OpenR2(c, b, sk_b, a, \mathsf{pk}(sk_a)) \end{array} \right]$$

By the same sequence of labelled transition as described previously, $\mathsf{out}(x, u_1)$ and $\mathsf{in}(c, e)$, we reach a network annotated with an additional sequent $\vdash e$. From that point we can access the following state where in addition $e \neq a$ is added to the constraint system, thereby systematising what we have described previously.

$$\text{new } sk_a, sk_b, m; C_0; \mathsf{pk}(sk_a), \mathsf{pk}(sk_b), \left\{ \left(a, \{m\}_{\mathsf{pk}(sk_b)} \right) \right\}_{\mathsf{pk}(sk_b)} \vdash e; e \neq a$$

$$\left[\left\{ \begin{array}{l} pk_a \mapsto \mathsf{pk}(sk_a), \\ pk_a \mapsto \mathsf{pk}(sk_b), \\ u_1 \mapsto \left\{ \left(a, \{m\}_{\mathsf{pk}(sk_b)} \right) \right\}_{\mathsf{pk}(sk_b)} \end{array} \right\} \middle| \begin{array}{l} \mathsf{in}(c, x); \\ \text{if } \mathsf{fst}(\mathsf{dec}(x, sk_a)) = b \text{ then} \\ \text{if } \mathsf{dec}(\mathsf{snd}(\mathsf{dec}(x, sk_a)), sk_a) = m \text{ then} \\ \mathsf{secret}(m) \mid \\ !I_{DY2}(c, a, sk_a, b, \mathsf{pk}(sk_b)) \mid \\ \text{if } e = a \text{ then} \\ \quad R_{DY2}(c, b, sk_b, a, pk_a) \\ \text{else} \\ \quad \mathsf{in}(c, pk_e); \\ \quad R_{DY2}(c, b, sk_b, e, pk_e) \\ \mid !OpenR2(c, b, sk_b, a, \mathsf{pk}(sk_a)) \end{array} \right]$$

Step 4. We now resume our state exploration where states are now extended processes enhanced with a constraint system.

From the above state, since we have assumed that $e \neq a$, we can perform transitions labelled with $\mathsf{in}(c, pk_e)$, and then $\mathsf{in}(c, y)$, to reach the following state, with respect to an updated new constraint system C_1.

4.5 Calculating input labels using the sequent calculus

$$\text{new } sk_a, sk_b, m; C_1 \left[\begin{array}{l} \left\{ \begin{array}{l} pk_a \mapsto \text{pk}(sk_a), \\ pk_a \mapsto \text{pk}(sk_b), \\ u_1 \mapsto \left\{ \left(a, \{m\}_{\text{pk}(sk_b)} \right) \right\}_{\text{pk}(sk_b)} \end{array} \right\} \left| \begin{array}{l} \text{in}(c, x); \\ \text{if } \text{fst}(\text{dec}(x, sk_a)) = b \text{ then} \\ \text{if } \text{dec}(\text{snd}(\text{dec}(x, sk_a)), sk_a) = m \text{ then} \\ \text{secret}(m) \mid \\ !I_{DY2}(c, a, sk_a, b, \text{pk}(sk_b)) \mid \\ \text{if } \text{fst}(\text{dec}(y, sk_b)) = e \text{ then} \\ \text{let } m = \text{dec}(\text{snd}(\text{dec}(y, sk_b)), sk_b) \text{ in} \\ \text{out}\left(c, \left\{ \left(b, \{m\}_{pk_e} \right) \right\}_{pk_e} \right) \mid \\ !OpenR2(c, b, sk_b, a, \text{pk}(sk_a)) \end{array} \right. \end{array} \right]$$

For each input, we generate constraints, expressed as sequents, that determine which outputs were available at the moment when the input occurred. Thus the updated constraints C_1 is as follows.

$$C_0; e \neq a;$$
$$\text{pk}(sk_a), \text{pk}(sk_b), \left\{ \left(a, \{m\}_{\text{pk}(sk_b)} \right) \right\}_{\text{pk}(sk_b)} \vdash e \; ;$$
$$\text{pk}(sk_a), \text{pk}(sk_b), \left\{ \left(a, \{m\}_{\text{pk}(sk_b)} \right) \right\}_{\text{pk}(sk_b)} \vdash pk_e \; ;$$
$$\text{pk}(sk_a), \text{pk}(sk_b), \left\{ \left(a, \{m\}_{\text{pk}(sk_b)} \right) \right\}_{\text{pk}(sk_b)} \vdash y$$

Notice that constraint system new $sk_a, sk_b, m; C_1$ has a solution. To see why, observe that each sequent is provable with respect to fresh names sk_a, sk_b, m, since the rule SOL can be immediately applied. Observe that when input variables are selected, as above, to be fresh with respect to what has been seen before, a solvable constraint system will always be extended to another solvable constraint system by performing such an input (by using the O-IN-ALIAS rule in Fig. 4.6). Furthermore, such fresh input variables are the most general forms of inputs thereby ranging over all possible input values.

Step 3. We try to perform the output of the honest responder, but initially we cannot since the guard of the responder that received the above inputs e, pk_e and y is not yet enabled. We say, "is not yet enabled," since this does not mean that the guard does not hold. What it does mean is that we do not yet have enough information about the inputs to decide whether the guard $\text{fst}(\text{dec}(y, sk_b)) = e$ is true.

To address this issue, we appeal to our concept of one constraint system accessing another via a substitution from Def. 4.1. When accessibility holds via a substitution, the substitution is applied to the extended process also, instantiating the variables that are changed everywhere. Accessibility, by definition, ensures that the constraints remain solvable, which guarantees that there is at least one possible choice of message for each input that could actually have led to the state accessed.

To determine the substitutions required we look at the guard that should be hold for a transition to be enabled. The guard $\text{fst}(\text{dec}(y, sk_b)) = e$ holds only if, from the perspective of the honest thread, y is of the form $\{(e, y_1)\}_{\text{pk}(sk_b)}$ for some fresh variable y_1. This leads us to considering the following two options.

1. Since the attacker knows the public key of b, via the alias pk_b, the attacker has the option to build a message passing the guards directly. This can be achieved by applying a substitution of the form $\{y \mapsto \{(e, y_1)\}_{\text{pk}(sk_b)}\}$ to the entire state and system of constraints. This leads to the following constraint, which is provable using the axioms and introduction rules (we invite the reader to fill the missing proof tree indicated by dots).

$$\cfrac{\cfrac{\vdots}{\text{new } sk_a, sk_b, m; \text{pk}(sk_a), \text{pk}(sk_b), \left\{\left(a, \{m\}_{\text{pk}(sk_b)}\right)\right\}_{\text{pk}(sk_b)} \vdash \text{pk}(sk_b)} \text{ Ax}}{\text{new } sk_a, sk_b, m; \text{pk}(sk_a), \text{pk}(sk_b), \left\{\left(a, \{m\}_{\text{pk}(sk_b)}\right)\right\}_{\text{pk}(sk_b)} \vdash \{(e, y_1)\}_{\text{pk}(sk_b)}} \text{ I-ENC}$$

The existence of a solution for this input constraint means that we are permitted to apply an input transition labelled $\text{in}\left(c, \{(e, y_1)\}_{pk_b}\right)$ at the previous step, rather than the more general input transition $\text{in}(c, y)$, where the alias for the public key is used at the point the axiom is applied in the derivation. Observe that we have left open the interpretation of fresh variable y_1 and also the fake identity assumed by the attacker e.

2. Alternatively, substitution $\left\{y \mapsto \left\{\left(a, \{m\}_{\text{pk}(sk_b)}\right)\right\}_{\text{pk}(sk_b)}, e \mapsto a, pk_e \mapsto pk_a\right\}$ can be applied, i.e., the attacker can assume the identity of a and use exactly the message that was sent. This does result in the following provable constraint.

$$\cfrac{}{\text{new } sk_a, sk_b, m; \text{pk}(sk_a), \text{pk}(sk_b), \left\{\left(a, \{m\}_{\text{pk}(sk_b)}\right)\right\}_{\text{pk}(sk_b)} \vdash \left\{\left(a, \{m\}_{\text{pk}(sk_b)}\right)\right\}_{\text{pk}(sk_b)}} \text{ Ax}$$

Notice that the existence of a solution to the above sequent corresponds to a particular input. In particular, it tells us that the input variable y at the previous step that generated the constraint should be labelled instead with $\text{in}(c, u_1)$, since when we apply the active substitution to u_1 we get the message for y. That is, u_1 is the recipe for describing $\left\{\left(a, \{m\}_{\text{pk}(sk_b)}\right)\right\}_{\text{pk}(sk_b)}$ from the perspective of the attacker. Remember that the attacker can only use free variables and never private names such as m and sk_b directly.

This second option however can be discarded, since the resulting constraint system is not solvable. In particular, the inequality $e \neq a$ is violated.

In this case, accessibility (Def. 4.1) tells us to consider the first of the above two substitutions going forward. In summary, there is a substitution sufficient to enable an equality in a guard, that when applied to the current constraints C_1 results in a solvable constraint system, and hence the substitution can be applied to the state to access the next state to consider, as given below.

4.5 Calculating input labels using the sequent calculus

$$\dfrac{\text{new } sk_a, sk_b, m; C_1\{y \mapsto \{(e, y_1)\}_{\text{pk}(sk_b)}\} \left\{ \begin{array}{l} pk_a \mapsto \text{pk}(sk_a), \\ pk_a \mapsto \text{pk}(sk_b), \\ u_1 \mapsto \left\{\left(a, \{m\}_{\text{pk}(sk_b)}\right)\right\}_{\text{pk}(sk_b)} \end{array} \right\}}{\begin{array}{l} \text{in}(c, x); \\ \text{if } \text{fst}(\text{dec}(x, sk_a)) = b \text{ then} \\ \text{if } \text{dec}(\text{snd}(\text{dec}(x, sk_a)), sk_a) = m \text{ then} \\ \text{secret}(m) \mid \\ !I_{DY2}(c, a, sk_a, b, \text{pk}(sk_b)) \mid \\ \text{if } \text{fst}\left(\text{dec}\left(\{(e, y_1)\}_{\text{pk}(sk_b)}, sk_b\right)\right) = e \text{ then} \\ \text{let } m = \text{dec}\left(\text{snd}\left(\text{dec}\left(\{(e, y_1)\}_{\text{pk}(sk_b)}, sk_b\right)\right), sk_b\right) \text{ in} \\ \text{out}\left(c, \left\{\left(b, \{m\}_{pk_e}\right)\right\}_{pk_e}\right) \mid \\ !OpenR2(c, b, sk_b, a, \text{pk}(sk_a)) \end{array}}$$

Step 5. Since, after applying the substitution $\{y \mapsto \{(e, y_1)\}_{\text{pk}(sk_b)}\}$, the guard $\text{fst}\left(\text{dec}\left(\{(e, y_1)\}_{\text{pk}(sk_b)}, sk_b\right)\right) = e$ before the output of the responder is enabled, an output action $\text{out}(c, u_2)$ can be performed leading to the following state.

$$\text{new } sk_a, sk_b, m; C_1\{y \mapsto \{(e, y_1)\}_{\text{pk}(sk_b)}\}\left[\begin{array}{l|l} \left\{\begin{array}{l} pk_a \mapsto \text{pk}(sk_a), \\ pk_a \mapsto \text{pk}(sk_b), \\ u_1 \mapsto \left\{\left(a, \{m\}_{\text{pk}(sk_b)}\right)\right\}_{\text{pk}(sk_b)}, \\ u_2 \mapsto \left\{\left(b, \{\text{dec}(y_1, sk_b)\}_{pk_e}\right)\right\}_{pk_e} \end{array}\right\} & \begin{array}{l} \text{in}(c, x); \\ \text{if } \text{fst}(\text{dec}(x, sk_a)) = b \text{ then} \\ \text{if } \text{dec}(\text{snd}(\text{dec}(x, sk_a)), sk_a) = m \text{ then} \\ \text{secret}(m) \mid \\ !I_{DY2}(c, a, sk_a, b, \text{pk}(sk_b)) \mid \\ 0 \mid \\ !OpenR2(c, b, sk_b, a, \text{pk}(sk_a)) \end{array} \end{array} \right]$$

Observe that, in the above process, the responder has now completed all its actions, as indicated by the presence of the terminated process 0. Also, observe that the most recent output is recorded using alias u_2 in the active substitution (the message output has been reduced using the equational theory for readability).

Step 6. At this point, there are several choices. The attacker can try to reach the secret event in the honest initiator thread by performing an input for the responder, by finding permitted values for the input that passes the match guard and then calculating a way to derive the secret. This heuristic often works, e.g., this way we can calculate the attack on the Dolev-Yao 1 and Needham-Schoeder protocols. However, for the Dolev-Yao 2 cascade protocol, the fact that the public key is used twice in messages, in a nested fashion, suggest a more sophisticated strategy for the attacker.

To proceed with finding an attack on the Dolev-Yao 2, the attacker can try to open a second session with the honest responder. That is, we repeat the above two steps, where the attacker creates a fresh identity to connect to the responder b and performs some input. The difference this time is that the attacker may make use of the output

u_1 produced in the previous steps. Hence, by following similar reasoning as above we eventually reach a state below, after performing the actions $\mathsf{in}(c, e')$, $\mathsf{in}(c, pk_{e'})$, $\mathsf{in}(c, z)$, and $\mathsf{out}(c, u_3)$.

$$\mathsf{new}\ sk_a, sk_b, m; C_2 \left[\left\{ \begin{pmatrix} pk_a \mapsto \mathsf{pk}(sk_a), \\ pk_a \mapsto \mathsf{pk}(sk_b), \\ u_1 \mapsto \left\{ \left(a, \{m\}_{\mathsf{pk}(sk_b)}\right)\right\}_{\mathsf{pk}(sk_b)}, \\ u_2 \mapsto \left\{ \left(b, \{\mathsf{dec}(y_1, sk_b)\}_{pk_e}\right)\right\}_{pk_e}, \\ u_3 \mapsto \left\{ \left(b, \{\mathsf{dec}(y_2, sk_b)\}_{pk_{e'}}\right)\right\}_{pk_{e'}} \end{pmatrix} \right\} \middle| \begin{array}{l} \mathsf{in}(c, x); \\ \mathsf{if}\ \mathsf{fst}(\mathsf{dec}(x, sk_a)) = b\ \mathsf{then} \\ \mathsf{if}\ \mathsf{dec}(\mathsf{snd}(\mathsf{dec}(x, sk_a)), sk_a) = m\ \mathsf{then} \\ \mathsf{secret}(m)\ | \\ !I_{DY2}(c, a, sk_a, b, \mathsf{pk}(sk_b))\ | \\ 0\ |\ 0\ | \\ !OpenR2(c, b, sk_b, a, \mathsf{pk}(sk_a)) \end{array} \right]$$

The constraint system C_2 accumulated at this point includes the following sequents, both of which are provable, which is necessary in order to be allowed to reach the given state, all with respect to fresh names sk_a, sk_b, m.

$$\mathsf{pk}(sk_a), \mathsf{pk}(sk_b), \left\{\left(a, \{m\}_{\mathsf{pk}(sk_b)}\right)\right\}_{\mathsf{pk}(sk_b)} \vdash \{(e, y_1)\}_{\mathsf{pk}(sk_b)}$$

$$\mathsf{pk}(sk_a), \mathsf{pk}(sk_b), \left\{\left(a, \{m\}_{\mathsf{pk}(sk_b)}\right)\right\}_{\mathsf{pk}(sk_b)}, \left\{\left(b, \{\mathsf{dec}(y_1, sk_b)\}_{pk_e}\right)\right\}_{pk_e} \vdash \{(e', y_2)\}_{\mathsf{pk}(sk_b)}$$

Step 7. We now proceed by sending an input to the honest initiator, by using the action $\mathsf{in}(c, w)$, where w is a fresh variable. The input results in the following sequent being added to the above system of constraints.

$$\mathsf{pk}(sk_a), \mathsf{pk}(sk_b), \left\{\left(a, \{m\}_{\mathsf{pk}(sk_b)}\right)\right\}_{\mathsf{pk}(sk_b)}, \left\{\left(b, \{\mathsf{dec}(y_1, sk_b)\}_{pk_e}\right)\right\}_{pk_e}, \left\{\left(b, \{\mathsf{dec}(y_2, sk_b)\}_{pk_{e'}}\right)\right\}_{pk_{e'}} \vdash w$$

The state at this point is as follows, which contains the free variable w, representing all possible inputs.

$$\mathsf{new}\ sk_a, sk_b, m; C_2 \left[\left\{ \begin{pmatrix} pk_a \mapsto \mathsf{pk}(sk_a), \\ pk_a \mapsto \mathsf{pk}(sk_b), \\ u_1 \mapsto \left\{ \left(a, \{m\}_{\mathsf{pk}(sk_b)}\right)\right\}_{\mathsf{pk}(sk_b)}, \\ u_2 \mapsto \left\{ \left(b, \{\mathsf{dec}(y_1, sk_b)\}_{pk_e}\right)\right\}_{pk_e}, \\ u_3 \mapsto \left\{ \left(b, \{\mathsf{dec}(y_2, sk_b)\}_{pk_{e'}}\right)\right\}_{pk_{e'}} \end{pmatrix} \right\} \middle| \begin{array}{l} \mathsf{if}\ \mathsf{fst}(\mathsf{dec}(w, sk_a)) = b\ \mathsf{then} \\ \mathsf{if}\ \mathsf{dec}(\mathsf{snd}(\mathsf{dec}(w, sk_a)), sk_a) = m\ \mathsf{then} \\ \mathsf{secret}(m)\ | \\ !I_{DY2}(c, a, sk_a, b, \mathsf{pk}(sk_b))\ | \\ 0\ |\ 0\ | \\ !OpenR2(c, b, sk_b, a, \mathsf{pk}(sk_a)) \end{array} \right]$$

Step 8. Finally, we aim to determine whether there is a choice of free variables w, y_1 and y_2 which enables the secret action to occur. As you can see from the state above, this means we need to enable two guards which impose further restrictions on w and only if these hold do we attempt to produce the message m.

To enable the guards, we seek substitutions such that the following equalities are satisfied.

$$\mathsf{fst}(\mathsf{dec}(w, sk_a)) = b \quad \text{and} \quad \mathsf{dec}(\mathsf{snd}(\mathsf{dec}(w, sk_a)), sk_a) = m$$

4.5 Calculating input labels using the sequent calculus

The above guards force w, from the perspective of the honest agent, to take on a value of the form $\left\{\left(b, \{m\}_{\text{pk}(sk_a)}\right)\right\}_{\text{pk}(sk_a)}$. To know whether this is possible we seek substitutions, from the perspective of the attacker, under which the above constraints hold. So, to summarise, we aim to find a solution, i.e., a substitution for the free variables such that following constraints are provable, with respect to fresh names sk_a, sk_b, m. The final constraint below corresponds to the secrecy event, which, of course, should not occur if secrecy holds. We abbreviate $\text{pk}(sk_a), \text{pk}(sk_b), \left\{\left(a, \{m\}_{\text{pk}(sk_b)}\right)\right\}_{\text{pk}(sk_b)}$ with Γ_1 in the following constraints.

$$\Gamma_1 \vdash \{(e, y_1)\}_{\text{pk}(sk_b)}$$
$$\Gamma_1, \left\{\left(b, \{\text{dec}(y_1, sk_b)\}_{pk_e}\right)\right\}_{pk_e} \vdash \{(e', y_2)\}_{\text{pk}(sk_b)}$$
$$\Gamma_1, \left\{\left(b, \{\text{dec}(y_1, sk_b)\}_{pk_e}\right)\right\}_{pk_e}, \left\{\left(b, \{\text{dec}(y_2, sk_b)\}_{pk_{e'}}\right)\right\}_{pk_{e'}} \vdash \left\{\left(b, \{m\}_{\text{pk}(sk_a)}\right)\right\}_{\text{pk}(sk_a)}$$
$$\Gamma_1, \left\{\left(b, \{\text{dec}(y_1, sk_b)\}_{pk_e}\right)\right\}_{pk_e}, \left\{\left(b, \{\text{dec}(y_2, sk_b)\}_{pk_{e'}}\right)\right\}_{pk_{e'}} \vdash m$$

As mentioned previously, an efficient way to find proof of such sequents is to first apply all elimination rules. Applying elimination rules breaks down the messages on the left of the turnstile.

In order to break down the messages encrypted with pk_e and $pk_{e'}$ (using the DEC rule), it is necessary to create fresh variables sk_e and $sk_{e'}$ and apply the substitution $\{pk_e \mapsto \text{pk}(sk_e), pk_{e'} \mapsto \text{pk}(sk_{e'})\}$. Observe that, up to this point, it was possible to delay this decision, leaving open the possibility of the attacker reusing the public key of an honest agent at the given point in the protocol (which does not help the attacker here, but may help for another protocol). Also, we can get rid of the deconstructors such as $\text{dec}(y_1, sk_b)$ by substituting the message y_1 with the most general form of message that would be decrypted by the key. The relevant substitution, we can apply here is $\{y_1 \mapsto \{x_1\}_{\text{pk}(sk_b)}, y_2 \mapsto \{x_2\}_{\text{pk}(sk_b)}\}$, after which we can reduce the deconstructor using the equational theory. So if we apply, for example, the substitution to $\text{dec}(y_1, sk_b)$ and reduce the decryption then we obtain y_1. We apply the above substitutions and hence enable the elimination rules to break down the sequents and obtain the following constraints.

$$\Gamma_1 \vdash \left\{\left(e, \{x_1\}_{\text{pk}(sk_b)}\right)\right\}_{\text{pk}(sk_b)}$$
$$\Gamma_1, b, x_1 \vdash \left\{\left(e', \{x_2\}_{\text{pk}(sk_b)}\right)\right\}_{\text{pk}(sk_b)}$$
$$\Gamma_1, b, x_1, b, x_2 \vdash \left\{\left(b, \{m\}_{\text{pk}(sk_a)}\right)\right\}_{\text{pk}(sk_a)}$$
$$\Gamma_1, b, x_1, b, x_2 \vdash m$$

Let us elaborate a bit more before we continue. To be really precise, in addition, we also have constraints such as $\vdash a$ and $\vdash b$, that are generated right at the beginning as our initial constraint system based on the free variables in the term. They simply constrain these variables such that they are never instantiated with new knowledge accumulated during the execution, e.g., by unifying a with a secret, which would clearly not have been possible, since the identities of agents were determined before the protocol started execution. Also we have constraints on e, sk_e, e' and $sk_{e'}$, ensuring they can only be generated by the appropriate attacker knowledge, and the inequalities $a \neq e$ and $a \neq e'$, that were required to resolve mismatch. These simpler constraints are never violated in what follows, since we never change these variables.

Now, we ask what are the choices for x_1 and x_2 that make the above constraints provable, with respect to fresh names sk_a, sk_b, m. Consider the first constraint; there are two options:

1. x_1 is anything derived from the knowledge using introduction rules such that

$$\text{new } sk_a, sk_b, m; \text{pk}(sk_a), \text{pk}(sk_b), \left\{\left(a, \{m\}_{\text{pk}(sk_b)}\right)\right\}_{\text{pk}(sk_b)} \vdash x_1$$

   Pursuing this further does not lead to an attack.

2. $\left\{x_1 \mapsto \left(a, \{m\}_{\text{pk}(sk_b)}\right)\right\}$, which allows the following proof below to be established. In the following, as before, we write Γ_1 as an abbreviation for $\text{pk}(sk_a), \text{pk}(sk_b), \left\{\left(a, \{m\}_{\text{pk}(sk_b)}\right)\right\}_{\text{pk}(sk_b)}$ and all sequents are with respect to fresh names sk_a, sk_b, m.

$$\cfrac{\cfrac{\overline{\Gamma_1 \vdash e}\ \text{Sol} \quad \overline{\Gamma_1 \vdash \left\{\left(a, \{m\}_{\text{pk}(sk_b)}\right)\right\}_{\text{pk}(sk_b)}}\ \text{Ax}}{\Gamma_1 \vdash \left(e, \left\{\left(a, \{m\}_{\text{pk}(sk_b)}\right)\right\}_{\text{pk}(sk_b)}\right)}\ \text{I-PAIR} \quad \overline{\Gamma_1 \vdash \text{pk}(sk_b)}\ \text{Ax}}{\Gamma_1 \vdash \left\{\left(e, \left\{\left(a, \{m\}_{\text{pk}(sk_b)}\right)\right\}_{\text{pk}(sk_b)}\right)\right\}_{\text{pk}(sk_b)}}\ \text{I-ENC}$$

We move on to the next sequent, shown below, to which the substitution $\left\{x_1 \mapsto \left(a, \{m\}_{\text{pk}(sk_b)}\right)\right\}$ has also been applied.

$$\text{new } sk_a, sk_b, m; \Gamma_1, b, \left(a, \{m\}_{\text{pk}(sk_b)}\right) \vdash \left\{\left(e', \{x_2\}_{\text{pk}(sk_b)}\right)\right\}_{\text{pk}(sk_b)}$$

Observe, after applying the above substitution to x_1, we can apply new instances elimination rules to break down the pair of messages. This gives us a new possibility for instantiating x_2, in addition to those options we had for x_1. In particular, we can apply the substitution $\{x_2 \mapsto m\}$ leading to the following proof, all with respect to fresh names sk_a, sk_b, m.

4.5 Calculating input labels using the sequent calculus

$$\cfrac{\cfrac{\Gamma_1, a, \{m\}_{\mathsf{pk}(sk_b)} \vdash e'}{\Gamma_1, a, \{m\}_{\mathsf{pk}(sk_b)} \vdash \left(e', \{m\}_{\mathsf{pk}(sk_b)}\right)}\text{Sol} \quad \cfrac{\Gamma_1, a, \{m\}_{\mathsf{pk}(sk_b)} \vdash \{m\}_{\mathsf{pk}(sk_b)}}{}\text{Ax}}{\cfrac{\Gamma_1, a, \{m\}_{\mathsf{pk}(sk_b)} \vdash \left\{\left(e', \{m\}_{\mathsf{pk}(sk_b)}\right)\right\}_{\mathsf{pk}(sk_b)}}{\Gamma_1, \left(a, \{m\}_{\mathsf{pk}(sk_b)}\right) \vdash \left\{\left(e', \{m\}_{\mathsf{pk}(sk_b)}\right)\right\}_{\mathsf{pk}(sk_b)}}\text{E-PAIR}} \text{I-PAIR} \quad \cfrac{\Gamma_1, a, \{m\}_{\mathsf{pk}(sk_b)} \vdash \mathsf{pk}(sk_b)}{}\text{Ax}}\text{I-ENC}$$

Now, the final two constraints are relatively easy to solve. Applying the both substitutions from above to the remaining constraints allows also the third constraint to be proven, by using introduction rules and axioms, as shown below. In the following, Γ_2 is the messages $\Gamma_1, \left(a, \{m\}_{\mathsf{pk}(sk_b)}\right), b, m$, all with respect to fresh names sk_a, sk_b, m.

$$\cfrac{\cfrac{\Gamma_2 \vdash b}{}\text{Sol} \quad \cfrac{\cfrac{\Gamma_2 \vdash m}{}\text{Ax} \quad \cfrac{\Gamma_2 \vdash \mathsf{pk}(sk_a)}{}\text{Ax}}{\Gamma_2 \vdash b\{m\}_{\mathsf{pk}(sk_a)}}\text{I-ENC}}{\cfrac{\Gamma_2 \vdash \left(b, \{m\}_{\mathsf{pk}(sk_a)}\right)}{\Gamma_2 \vdash \left\{\left(b, \{m\}_{\mathsf{pk}(sk_a)}\right)\right\}_{\mathsf{pk}(sk_a)}}\text{I-PAIR} \quad \cfrac{\Gamma_2 \vdash \mathsf{pk}(sk_a)}{}\text{Ax}}{}}\text{I-ENC}$$

The final sequent is provable, since the variable x_2 has been mapped to m, which enables an axiom.

$$\cfrac{}{\mathsf{new}\ sk_a, sk_b, m;\ \Gamma_1, b, \left(a, \{m\}_{\mathsf{pk}(sk_b)}\right), b, m \vdash m}\text{Ax}$$

Thereby, we have found a solution to the constraints. That solution shows that we can enable the secret action of the honest initiator. There exists a substitution for the input variables, using only information available to the attacker at the point of input, that enables a sequence of transitions leading to the secret action. More specifically, the secret action below is enabled.

$$\mathsf{secret}(\mathsf{dec}(\mathsf{snd}(\mathsf{dec}(u_3, sk_{e'})), sk_{e'}))$$

The message $\mathsf{dec}(\mathsf{snd}(\mathsf{dec}(u_3, sk_{e'})), sk_{e'})$ is the way the attacker obtained the message m, which cannot be seen directly by the attacker, as modelled by the fresh variable binder. This message can be obtained automatically, by looking at the rules of the sequent calculus which were used to produce this message. Observe that three elimination rules were applied to obtain m, which correspond to applying the three deconstructors, as shown in the message in the secret action.

Summary. The attack trace discovered above is summarised by the following formula, which is satisfied by the state State_{DY2} defined at the beginning of this case study. The trace contains a secret event, so can be used as a proof certificate, providing efficiently checkable evidence that secrecy is violated.

$State_{DY2} \models$
$\langle \text{out}(c, u_1) \rangle$ Output by a.
$\langle \text{in}(c, e) \rangle$ Connect to b with e.
$\langle \text{in}(c, \text{pk}(sk_e)) \rangle$ with public key $\text{pk}(sk_e)$.
$\langle \text{in}(c, \{(e, u_1)\}_{pk_b}) \rangle$ Construct input for b.
$\langle \text{out}(c, u_2) \rangle$ Response from b.
$\langle \text{in}(c, e') \rangle$ Connect to b with e',
$\langle \text{in}(c, \text{pk}(sk_{e'})) \rangle$ with public key $\text{pk}(sk_{e'})$.
$\langle \text{in}(c, \{(e', \text{snd}(\text{dec}(\text{snd}(\text{dec}(u_2, sk_e)), sk_e)))\}_{pk_b}) \rangle$ Construct input for b.
$\langle \text{out}(c, u_3) \rangle$ Second response from b.
$\langle \text{in}(c, \{(b, \{\text{dec}(\text{snd}(\text{dec}(u_3, sk_{e'})), sk_{e'})\}_{pk_a})\}_{pk_a}) \rangle$ Trick a into completing.
$\langle \text{secret}(\text{dec}(\text{snd}(\text{dec}(u_3, sk_{e'})), sk_{e'})) \rangle$ Attacker observes secret.
true

Each of the messages in the above trace corresponds to the steps taken in the proof of the sequent producing the input or the secret message. The above attack trace describes formally the same attack that is depicted informally in the message sequence chart in Fig. 4.7.

This concludes our case study of using sequents to discover an attack on the Dolev-Yao 3 "cascade" protocol. In reflection, discovering the above attack may be rather difficult without applying a methodology. Either we can employ a methodology such as the one we have just applied or use our threat modelling knowledge and employ a tool to do such a calculation for us. We should not be reliant on having prior expertise on discovering attack patterns on cascade protocols. Thus using symbolic verification (manually or using a tool) is a better approach to this problem.

4.5.2 Practice in discovering attacks using sequents

We have already mentioned the Needham-Schroeder protocol, modelled the roles and stated that there is an attack. We can now apply the methodology from this section in order to discover an attack on the Needham-Schroeder protocol systematically. Since there are three message exchanges and both roles of the protocol produce an output based on an input, we must allow open connections for both roles, as we did for the Dolev-Yao protocol. That is, to model the "open network" threat model we should explicitly allow both roles to be instantiated by honest agents connecting to anyone, including the attacker. Thus we aim to verify the protocol under assumptions of a typical modern network, such as the internet or an office LAN, where connections across the network from compromised positions are possible.

We need the following threads, adapted from the exercises in Sec. 3.2.5. Suppose that $I_{NS}^{honest}(c, i, sk_i, r, pk_r)$ is the thread of the initiator role as you have already modelled in a previous exercise and $R_{NS}^{honest}(c, r, sk_r, i, pk_i)$ is the thread of the

4.5 Calculating input labels using the sequent calculus   117

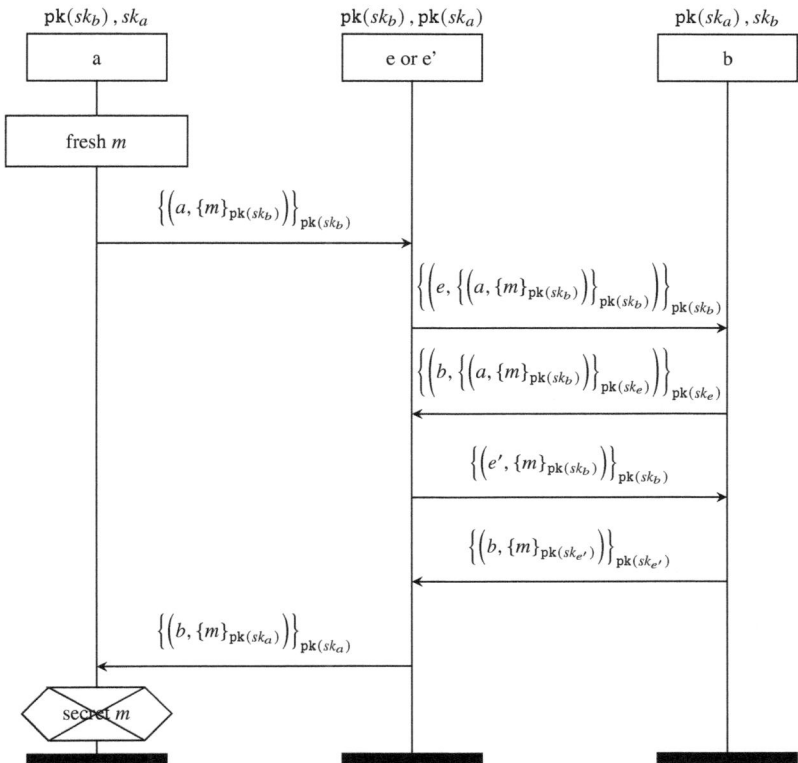

Fig. 4.7 Attack on the Dolev-Yao 2 protocol.

responder role. We mark them as honest, since you should have included the secrecy claims in each role. Now, take each of these roles and create another version where we remove the secrecy claims (as in the model of the responder in the Dolev-Yao protocols where the thread ends with 0), and call each of these modified threads $I_{NS}^{open}(c, i, sk_i, r, pk_r)$ and $I_{NS}^{open}(c, r, sk_r, i, pk_i)$. These modified threads are the open initiator and open responder respectively, and reflects that we do not assert secrecy when we connect to an agent we do not trust.

We now define an initial configuration, capturing a suitable "open network" threat model for the Needham-Schroeder protocol. To simplify things, since we know[3] there is an attack on secrecy from the perspective of the Responder, and that this attack requires an open Initiator. Therefore, we can safely simplify the Responder as shown

---

[3] This simplification of the threat model is permitted by magically knowing already the capabilities required to find an attack on this protocol. In general, you should check a protocol with both agents being open, not just the initiator.

below and still find an attack.

new sk_a, sk_b;	Declare the private keys of trusted agents a and b.
out($keys$, pk(sk_a));	Publish the public key of a.
out($keys$, pk(sk_b)); (	Publish the public key of b.
!$R_{NS}^{honest}(c, b, sk_b, a, \text{pk}(sk_a))$	Agent b responds to a session with agent a.
|	
!in(c, e);	Agent a initiates a session with anyone.
if $e = b$ then	
$\quad I_{NS}^{honest}(c, a, sk_a, b, \text{pk}(sk_b))$	
else	
$\quad$ in(c, pk_e);	
$\quad I_{NS}^{open}(c, a, sk_a, e, pk_e)$)	

Exercise 4.3 Apply the methodology we have developed so far to find an attack on secrecy from the perspective of an honest responder role talking to an honest initiator. That is:

1. Execute the protocol according to a heuristic. Since we aim to show that secrecy is broken for the honest responder b, this means that we should aim to start an honest session where b is awaiting an input from a. To do so, we should start an initiator session with a; but we try to trick a into talking to the attacker via a fake identity e which is used to connect to the open initiator session offered by a.
2. For outputs, we simply extend the state. For inputs, we should also generate sequents whenever they occur, where the sequents represent the outputs known to the attacker at the point of input.
3. We try to reach the secret event at the end of the responder role. In doing so we generate a system of constraints to solve. If there is a substitution that makes them provable then we have found the attack.
4. Write down the trace and message sequence chart for your attack.

Explain how Lowe proposed to fix the Needham-Schroeder protocol and explain why his fix prevents your attack. This requires you to research the literature in order to find the Needham-Schroeder-Lowe protocol.

4.6 Presenting proofs that secrecy properties hold

In this section, we evaluate a third version of the Dolev-Yao protocol (DY_3 in Fig. 4.4 and 4.8), and prove that the secrecy statement in this new version holds.

For the DY_1 and DY_2 protocols we have shown that secrecy does not hold. We proved this property by showing that an attack *exists*. However, if we have to prove that the secrecy property holds, we should prove that no attack exists, that is, we should prove that *for all* sequences of transitions we never reveal a secret. In order to do so, we show that there is a space of knowledge closed under all, infinitely many,

4.6 Presenting proofs that secrecy properties hold

$I_{DY3}(c, i, sk_i, r, pk_r) \triangleq$
new m;
$\text{out}\left(c, \{(i, m)\}_{pk_r}\right)$;
$\text{in}(c, x)$;
if $\text{fst}(\text{dec}(x, sk_i)) = r$ then
if $\text{snd}(\text{dec}(x, sk_i)) = m$ then
$\text{secret}(m)$

$R_{DY3}(c, r, sk_r, i, pk_i) \triangleq$
$\text{in}(c, x)$;
if $\text{fst}(\text{dec}(x, sk_r)) = i$ then
let $m = \text{snd}(\text{dec}(x, sk_r))$ in
$\text{out}\left(c, \{(r, m)\}_{pk(sk_i)}\right)$;

Fig. 4.8 A specification of the Dolev-Yao 3 protocol.

transitions, and, furthermore, in no state covered is a secret transition enabled. This requires the construction of a finite representation of an infinite amount of knowledge.

To prove that the secrecy claim in DY_3 holds, we show that even if the attacker intercepts the messages exchanged between a and b, and even if the attacker can connect to b and uses these messages intercepted, then the attacker never generates enough knowledge to reveal the secret m.

Consider the following constraint system representing knowledge of an attacker, where $k \leq n$, which we name $\Gamma_{n,k}$.

new $sk_a, sk_b, m_1, \ldots, m_n; \text{pk}(sk_a), \text{pk}(sk_b), \{(a, m_1)\}_{\text{pk}(sk_b)}, \ldots \{(a, m_n)\}_{\text{pk}(sk_b)},$
$\{(b, m_1)\}_{\text{pk}(sk_a)}, \ldots \{(b, m_k)\}_{\text{pk}(sk_a)}$

We know revealing secret m_i corresponds to the proving sequent $\Gamma \vdash m_i$. There is no such proof, since no rule can be applied, and hence it is impossible that if we have only this knowledge that a secret is revealed.

We now have to show that this knowledge is closed under all transitions. If that is so, we call this an *invariant* of the protocol. Clearly, after outputting the two public keys we reach a sequent with the above knowledge where $n = k = 0$. Firstly, we list below the possible transitions that do not involve open connections with b:

- If a starts a new session with b, then a fresh name, say m_{n+1} is extruded and the message $\{(a, m_n)\}_{\text{pk}(sk_b)}$ is added to the knowledge. Hence clearly, the knowledge is $\Gamma_{n+1,k}$ which is the form of the invariant.
- If b starts a session with a, then he must induce an input of the following form in order to trigger an output.

$$\Gamma_{n,k} \vdash \{(a, x)\}_{\text{pk}(sk_a)}$$

There are precisely two solutions to this.

 – The first solution is where $x = m_i$, which corresponds to following the usual execution of the protocol. In this case, the output induced is of the form $\{(b, m_i)\}_{\text{pk}(sk_a)}$. Now, either $1 \leq i \leq k$ in which case no knowledge is gained and the knowledge $\Gamma_{n,k}$ remains the same, or, w.l.o.g., $i = k + 1 \leq n$ and hence we have knowledge $\Gamma_{n,k+1}$, which is also in the form of the knowledge invariant.

– The other solution is where $\Gamma_{n,k} \vdash x$, that is, the attacker built the message using their knowledge in $\Gamma_{n,k}$. The resulting output is then $\{(b,x)\}_{\text{pk}(sk_a)}$. The important observation is that we know that we also have the following.

$$\Gamma_{n,k} \vdash \{(b,x)\}_{\text{pk}(sk_a)}$$

This means that the attacker has not gained any knowledge, since they already knew how to construct the message output. Thus our knowledge is still represented by $\Gamma_{n,k}$.

The above reasoning does not complete the proof, since we should also consider open connections where the attacker connects to b. However we would like to analyse more precisely the claim that if $\Gamma_{n,k} \vdash \{(b,x)\}_{\text{pk}(sk_a)}$, then $\Gamma_{n,k}, \{(b,x)\}_{\text{pk}(sk_a)}$ represents the same knowledge as $\Gamma_{n,k}$. In order to argue formally, we introduce the CUT rule below.

$$\frac{\Gamma \vdash M \quad \Gamma, M \vdash N}{\Gamma \vdash N} \text{ CUT}$$

The CUT extends our sequent calculus from Sec. 4.4.1, and is, essentially, a powerful deduction rule. You may read it as follows: if M is a lemma we can establish assuming the knowledge in Γ, then we can first prove the lemma that we know $\Gamma \vdash M$, and then we can use that when proving of N under the same assumptions. This generalises concepts such as *modus ponens* that the reader may have encountered in any introduction to logic.

Now, by applying the cut rule, we can establish that for any message N that we can derive from $\Gamma_{n,k}, \{(b,x)\}_{\text{pk}(sk_a)}$, that is if we have $\Gamma_{n,k}, \{(b,x)\}_{\text{pk}(sk_a)} \vdash N$. Since we know that we can prove $\Gamma_{n,k} \vdash \{(b,x)\}_{\text{pk}(sk_a)}$, we can establish the two sequents in the premise of an instance of the CUT rule and hence we have that $\Gamma_{n,k} \vdash N$. Since we reasoned with respect to any N, this proves that $\Gamma_{n,k}$ and $\Gamma_{n,k}, \Gamma_{n,k}, \{(b,x)\}_{\text{pk}(sk_a)}$ establish the same knowledge in the presence of CUT.

The problem now is that it is more difficult to make calculations with the CUT rule. There are infinitely many ways we can come up with and intermediate lemma and hence there are infinitely many proof trees if we apply CUT.

The power of the CUT-FREE sequent calculus comes from the fact that we do not need deduction – that is, we do not need to come up with lemmas. We can simply apply the rules of the sequent calculus and establish the conclusion directly without appealing to lemmas. Fortunately, for any good sequent calculus the following theorem holds, which is called CUT-elimination in the literature.

Theorem 4.1 *If $\Gamma \vdash M$ using the rules of the sequent calculus with the CUT rule, then $\Gamma \vdash M$ using the CUT-FREE sequent calculus.*

Now we can appeal to the above CUT-elimination theorem to argue that, even in the CUT-FREE sequent calculus, $\Gamma_{n,k}$ and $\Gamma_{n,k}, \{(b,x)\}_{\text{pk}(sk_a)}$ represent equivalent knowledge.

We now return to our proof of the secrecy of DY_3.

- The only remaining way to induce a new output is for the attacker to connect to b with a fake identity and public key. When we connect, we input some identity say

4.6 Presenting proofs that secrecy properties hold

e and public key pk_e, which may only be generated using the current knowledge of the attacker $\Gamma_{n,k}$. That is, we assume we have constraints $\Gamma_{n,k} \vdash e$ and $\Gamma_{n,k} \vdash pk_e$, and also $e \neq a$. Also, to induce an output, the input must be a solution for the following constraint.

$$\Gamma_{n,k} \vdash \{(e,x)\}_{\mathrm{pk}(sk_a)}$$

Since $e \neq a$, this is only possible if we construct the message directly. That is, we have a proof of the following form.

$$\cfrac{\cfrac{\overline{\Gamma_{n,k} \vdash e}\ \text{Sol} \quad \overline{\Gamma_{n,k} \vdash x}\ \text{Sol}}{\Gamma_{n,k} \vdash (e,x)}\ \text{I-Enc} \quad \cfrac{}{\Gamma_{n,k} \vdash \mathrm{pk}(sk_a)}\ \text{Ax}}{\Gamma_{n,k} \vdash \{(e,x)\}_{\mathrm{pk}(sk_a)}}\ \text{I-Enc}$$

Notice this means that it must be the case that we have $\Gamma_{n,k} \vdash x$ holds. We now observe that the resulting output will be $\{(b,x)\}_{pk_e}$. However, this contains no new knowledge since the following is provable.

$$\Gamma_{n,k} \vdash \{(b,x)\}_{\mathrm{pk}(pk_e)}$$

Hence, via the cut-elimination argument explained previously, $\Gamma_{n,k}, \{(b,x)\}_{\mathrm{pk}(pk_e)}$ represents the same knowledge as $\Gamma_{n,k}$. Hence we have preserved the invariant.

Since we have now considered all transitions that induce an output, shown that they preserve our invariant, and that knowledge does not reveal a secret, we can conclude that it is impossible in any state the system can reach that it will have enough knowledge to reveal any of the secrets. This concludes our proof.

4.6.1 Practice in proving secrecy

Proving security properties of a protocol is typically more difficult than finding attacks. For a proof we need to construct a representation of an infinite set of knowledge that we prove is invariant under all infinitely many transitions, as we have done above in this section. Therefore this is an advanced question, that will none-the-less give you mastery of these protocol verification techniques.

Exercise 4.4 (Advanced) Recall the Needham-Schroeder-Lowe protocol that you found in Exercise 4.3, through research of the literature.

- Find a knowledge invariant for the Needham-Schroeder-Lowe protocol.
- Show that that your knowledge invariant is preserved under all transitions, and formulate your argument as a proof, as presenting above in this section for DY_3.

This is more challenging than the proof for DY_3 above, since more messages are exchanged, resulting in a larger knowledge invariant. None-the-less, the method is similar. We explore all solutions to a sequent representing any input that induces and output. We then use the sequent calculus again to check whether that knowledge follows from the existing knowledge. Otherwise, we extend the attacker's knowledge with messages of the form that are output.

4.6.2 Practice in advanced aspects of semantics

The advanced exercises and explanations in this section are designed to assist the reader wishing to understand further the computational dynamics of the semantics covered so far. Sometimes we have loosely referred to a substitution as a unifier. Most general unifiers are substitutions that are sufficient to consider when we aim for two specific terms to be made equal [18, 109]. Since, we take care not to instantiate variables representing fresh names with general message terms, the definition below takes into account also a set of names that should not be changed by unifiers.

Definition 4.2 (Most general unifiers) Let $mgu_E^x(M, N)$ be a set of substitutions such that:

- for all $\sigma \in mgu_E^x(M, N)$ we have $M\sigma =_E N\sigma$ and $\mathbf{x} \mathbin{\#} \mathrm{dom}(\sigma)$.
- for all θ such that $M\theta =_E N\theta$ and $\mathbf{x} \mathbin{\#} \mathrm{dom}(\theta)$ there exists $\sigma \in mgu(M, N)$ and ρ such that $\sigma \circ \rho = \theta$.
- there is no $\sigma, \sigma' \in mgu_E^x(M, N)$ such that for some ρ we have $\sigma \circ \rho = \sigma'$.

In the theories we work with in this book $mgu_E^x(M, N)$ is always finite, and usually a singleton. Considering most general unifiers is typically sufficient, since most properties of the system enhanced with constraints are monotonic with respect to substitutions, and hence hold with respect to other unifiers if they hold with respect the most general unifiers. By monotonicity with respect to substitutions we mean properties such as: if $M =_E N$ then $M\sigma =_E N\sigma$, and, in the transitions system enhanced with constraints, if $E \xrightarrow{\pi} E'$ then $E\sigma \xrightarrow{\pi\sigma} E'\sigma$.

Assume that new $\mathbf{x}; \Gamma \nvdash M$ holds whenever a search for a proof of new $\mathbf{x}; \Gamma \vdash M$ fails, which we can always calculate through a finite search thanks to the fact that proof search in the sequent calculus is finite. Simply applying all elimination rules to the max and then follow the syntax of the message on the right hand side until no rule applies (or the proof completes in which case new $\mathbf{x}; \Gamma \nvdash M$ does not hold).

We now have the following finitely checkable rules for determining whether an inequality holds.

Proposition 4.2 *We have* new $\mathbf{x}; C \models M \neq N$ *whenever, for all* $\sigma \in mgu_E^x(M, N)$, *either:*

- *for some* $\Gamma \vdash L$ *in* C *we have* new $\mathbf{x}; \Gamma\sigma \nvdash L\sigma$
- *or for some* $K \vdash L$ *in* C *we have* $K\sigma =_E L\sigma$.

Exercise 4.5 The following questions help understand some richer calculations involving inequalities.

1. Use the above rules to prove the following, with respect to message theory E:

4.6 Presenting proofs that secrecy properties hold

    a. $\text{new } m, k; \vdash n \models \{m\}_{\text{pk}(k)} \neq \{n\}_{\text{pk}(k)}$,
    b. $\{m\}_{\text{pk}(k)} \neq \{\text{fst}((n, \ell))\}_{\text{pk}(k)} \models (m, x) \neq (n, x)$,
    c. $\text{new } m, n; \models m \neq n$.

2. Notice that the O-EXTRUDE rule in Fig. 4.6 expands the names in the constraint system with the name extruded. This ensures that the extruded name is treated as a fresh name. To see this, consider the following state annotated with a constraint system.

$$\vdash y \, [id \,|\, \text{new } x; \text{if } x = y \text{ then } 0 \text{ else else out}(y, y)]$$

Provide the full derivation tree for the transition that the process above makes, paying attention to the use of O-EXTRUDE and O-MISMATCH.

3. Use the facts that we know about unifiers, Definition 4.1, and properties of unifiers and sequents discussed above to prove Proposition 4.2. This involves unfolding definitions and applying standard logical reasoning.

We tie up loose ends with an advanced observation connecting systems presented so far. When working with extended processes enhanced with a constraint system, as described in Sec. 4.5.1, we constrain fresh variables, such as pk_e in the following process with a sequent.

$$\vdash c; \vdash pk_e \left[id \,\bigg|\, \text{new } m; \text{out}\!\left(c, \{m\}_{pk_e}\right); \text{secret}(m) \right]$$

The constraint $\vdash pk_e$ ensures that the variable pk_e cannot be instantiated with knowledge output in the future. The constraint however does not prevent the free variables from being instantiated with messages formed using further free variables. For example, the system is still solvable after applying substitution $\{pk_e \mapsto \text{pk}(sk_e)\}$, since sk_e is a free variable and hence rules I-PK and SOL apply from Fig. 4.5.[4]

This observation leads to adjustments to Theorem 3.1, characterising secrecy, when we make use of constraint systems. We refer to a "lazy" trace, to be one as described in Sec. 4.5.1, where we start with a constraint system covering all open variables in a process and move between states using both labelled transitions and accessibility. Now for process P such that $\text{fv}(P) = \{x_0, \ldots x_n\}$, if, starting with $\vdash x_0; \ldots \vdash x_n \, [id \,|\, P]$, there exists a lazy sequence of transitions that reveals a secret, then there exists a substitution σ such that $[id \,|\, P\sigma]$ reveals a secret, as per Definition 3.2. This is a soundness property of reasoning with respect to lazy traces, accounting for a possible substitution, as illustrated above. Soundness can be proven by appealing repeatedly to a lemma: for every $D \xrightarrow{\pi} E$ in such a lazy trace, if E can access E', then there exists D' and σ such that D can access D' and $D' \xrightarrow{\pi\sigma} E'$. The lemma, essentially says, "lazy choices could have been made earlier," and applying the lemma repeatedly brings all the choices made lazily via accessibility right back to the points where messages were input and, finally, to where a simple substitution is applied to the initial state. The constructive content of such

---

[4] Encoding this example in ProVerif will not find this attack. However, exactly the same semantics can be achieved in ProVerif by prefixing process P with inputs parameterised on each $x \in \text{fn}(P)$, on a channel fresh for P.

a soundness proof is effectively what is appealed to when constructing the formula at the end of Sec. 4.5.1.

Soundness, above, establishes that if a trace is found making use of a constraint system, there is an attack on secrecy. The counterpart, completeness, ensures that if there is an attack on secrecy, then it can indeed be found making use of a constraint system. That is, completeness would establish that for every $[id \,|\, P\sigma] \vdash \langle \pi'_0 \rangle \langle \pi'_1 \rangle \ldots \langle \pi'_n \rangle \langle \text{secret}(M') \rangle$ there exists a lazy trace. This is relatively straightforward to see since, in Fig. 4.6 only O-IN-ALIAS adds a constraint to the system such that $\text{ran}(\theta) \vdash N\theta$. Since N is fresh for $\mathbf{x}, \mathbf{y}$ we can read off a proof of $\text{new}\,\mathbf{x}, \mathbf{y}; \text{ran}(\theta) \vdash N\theta$ from the structure of N, due to a tight correspondence between message terms such as N that do not refer to fresh names and proofs of $\text{new}\,\mathbf{x}, \mathbf{y}; \text{ran}(\theta) \vdash N\theta$.

The above two paragraphs, of course, do not constitute a formal proof of soundness and completeness. For example, the lemma hinted at above would need to be proven with respect to the derivation trees for labelled transitions. The purpose of this book is not to provide all meta theory, but the theory is presented accurately so that the machinery explained can be used in confidence.

Part III
Authentication and Privacy

Chapter 5
Authentication by agreement

Abstract The strong authentication property agreement, where honest agents agree on the content of all message exchanged, is introduced formally. Differences between injective and non-injective agreement are explained via a series of minimal examples. Expressing agreement using tools is also demonstrated. Differences in threat models due to public keys and symmetric keys are touched on. A large multiparty case study mixing public keys and symmetric keys based on OpenID Connect is covered in detail. Exercises lean towards the discovery of attacks and the evaluation of patches guided by secrecy and multiparty agreement properties.

5.1 Introduction

In the previous chapter we focused on reasoning about the *secrecy* of messages exchanged between the agents executing security protocols. Now we will consider other types of properties that secure protocols must satisfy in order to be useful for their ultimate goal, which can typically be to allow agents to securely share resources with their intended communication partners.

Authentication properties allow agents to assert claims about the identity of their communication partners, and the manner in which the exchange occurred. Strong authentication properties, such as *agreement*, ensure that all agents are certain that the communication occurred as prescribed, and with the intended communication partners. Weaker authentication properties can also be useful for assessing weakness in protocols, for example a failure of *recent aliveness* happens if an honest communicating partner need not be present during the execution of a protocol, using messages sent in the past for instance. Typically, authentication properties are combined with a requirement that secrets are not leaked to the adversary. In this chapter, we consider only the authentication property agreement. The rationale is that many tools are well equipped to model and verify this property, whereas authentication properties in later chapters pose more challenges for current tools.

5.2 Agreement – a strong authentication property

Strong authentication guarantees that honest communicating partners followed the protocol as intended, despite the best efforts of an attacker. The protocol may not be followed in some protocols for various reasons, including the attacker being able to covertly manipulate some messages exchanged, for instance changing transaction details without necessarily changing the identities involved. In Sec. 2.2 and 2.3, we already discussed informally such authentication properties and explained their importance for critical infrastructure such as ePassports and ePayments. In this section, we introduce formal definitions of agreement.

> *Authentication by agreement.* Agreement properties allow an agent to claim that honest agents participating in the protocol execution agree on the content of all messages exchanged. We cover two flavours of agreement:
>
> - *Agreement*: Agreement ensures that every time an agent believes they have reached an agreement with honest participants, there really were honest communicating partners that sent and received the same messages. A weakness of this definition is that, those messages could also have been used to authenticate another session of the protocol. As mentioned in Sec. 2.2, this opens the door for the adversary running *replay attacks*.
> - *Injective agreement*: Injective agreement strengthens the above to ensure that every time an agent executes the role containing the claim, the agreement occurs with a distinct session of every other agent involved.

These agreement properties are widely recognised in the security community [82]. If injectivity is explicitly not a requirement, the first of these properties is referred to as *non-injective agreement*. Non-injective agreement can be relevant in some protocols, for example if something generated by part of a protocol, e.g., a credential, may be reused across multiple sessions [36].

To study agreement we extend the syntax of processes to include a suitable claim.

P ::= ...   as before (Sec. 3.4)
  | agree(M); P   agreement claim   (EXTENDED SYNTAX OF PROCESSES)

Next, we introduce a number of tools that will help us analyse agreement claims. Firstly, we introduce transitions labelled with sequences of actions and the subsequence relation. We will use the subsequence relation to pick out a particular part of an execution.

Definition 5.1 (Traces and sub-traces) For a sequence of actions $tr = \pi_0, \pi_1, \ldots \pi_n$ we write $A \xRightarrow{tr} B$ whenever there exists transitions $A \xrightarrow{\pi_0} A_0$ and $A_i \xrightarrow{\pi_{i+1}} A_{i+1}$ for $0 \le i < n$ and $A_n = B$.

We define when a sequence of actions tr' is a subsequence of another sequence tr, written $tr' \sqsubseteq tr$, inductively as follows:

5.2 Agreement – a strong authentication property

- the base case is $\epsilon \sqsubseteq tr$, where ϵ is the empty sequence; and
- $\pi, tr' \sqsubseteq \pi', tr$ holds whenever $\pi = \pi'$ and $tr' \sqsubseteq tr$, or we have $\pi, tr' \sqsubseteq tr$.

We extend our satisfaction relation to express equalities as follows.

Definition 5.2 For a state State_A and message M and N, we define $\text{State}_A \models M = N$ according to the following rules.

$$\frac{M\sigma =_E N\sigma}{[\sigma \mid P] \models M = N} \qquad \frac{\text{State}_A \{x \mapsto z\} \models M = N \quad z \,\#\, \text{new}\, x; \text{State}_A, M, N}{\text{new}\, x; \text{State}_A \models M = N}$$

Inequality $\text{State}_B \models M \neq N$ holds whenever $\text{State}_B \models M = N$ fails, i.e., cannot be proven using the above rules.

In the above, the second rule ensures that the fresh names in the process never appear directly in the messages M and N being compared. This effect is achieved in a similar way to the EXTRUDE rule, where the bound variable may be renamed to avoid name capture. The first clause applies the active substitution to the message and checks whether they are equal according to the equational theory E. The process P plays no role in this definition.

The following is an example where an equality is satisfied in a state. Here, multiple rules removing freshness constraints are applied at once, and no renaming is required since there are no clashes between free and bound variables.

$$\frac{(c, hello) =_E \left(\text{dec}\left(\{c\}_{\text{pk}(k)}, k\right), hello\right) \qquad \left[\left\{\begin{array}{l} u \mapsto (c, hello), \\ v \mapsto \{c\}_{\text{pk}(k)}, w \mapsto k \end{array}\right\} \middle| 0\right] \models u = (\text{dec}(v, w), hello) \qquad c, k \,\#\, 0, u, (\text{dec}(v, w), hello)}{\text{new}\, c, k; \left[\left\{\begin{array}{l} u \mapsto (c, hello), \\ v \mapsto \{c\}_{\text{pk}(k)}, w \mapsto k \end{array}\right\} \middle| 0\right] \models u = (\text{dec}(v, w), hello)}$$

We now have the machinery to define agreement properties. The following defines agreement for a network where the agent authenticating all messages is acting on a single channel K, while all the communications that they wish to authenticate are on another single channel, say $f(K)$, where f is a function mapping channels of the agent authenticating to the channel of the agent being authenticated. If we insist that function is injective, we obtain injective agreement.

Definition 5.3 (agreement) A state State_A satisfies *agreement* whenever for all sequences of actions tr such that $\text{State}_A \xrightarrow{tr} \text{State}_B$, there exists a function f on messages such that the following holds.

For all $\pi_n, \ldots \pi_1, \pi_0, \text{agree}(K) \sqsubseteq tr$ such that $\pi_0 = \text{in}(K, M_0)$ and for $1 \leq i \leq n$, either $\pi_i = \text{in}(K, M_i)$ or $\pi_i = \text{out}(K, u_i)$, there exist actions $\kappa_n, \ldots \kappa_0$ such that $\kappa_n, \ldots \kappa_0 \sqsubseteq tr$ and, for all $0 \leq i \leq n$, we have:

- if $\pi_i = \text{in}(K, M_i)$ then $\kappa_i = \text{out}(L_i, u_i)$, and if $\pi_i = \text{out}(K, u_i)$ then $\kappa_i = \text{in}(L_i, M_i)$, and
- $\text{State}_B \models u_i = M_i$, and $\text{State}_B \models L_i = f(K)$.

If, in addition, f is injective in the above, then State_A satisfies *injective agreement*.

The above definition is the richest we have encountered, so we break it down. Firstly, notice we range over all traces tr and hence we cover all executions. Within each trace, we consider all patterns that we can match of a particular form $\pi_n, \ldots \pi_1, \pi_0, \text{agree}(K)$. That pattern ends with an agreement claim, and is proceeded by several input and output actions on the channel of the authenticator K. Notably the final action before authentication π_0 must be an input. This is because we have no control over whether or not anything output at the end of the protocol will reach its destination, without further inputs, and hence we cannot include such outputs after that final input. Notice that it is implicit that we should consider a maximal sequence of inputs and outputs on the given channel, since we insist on ranging over all such sequences, including the one containing maximal sequences of inputs and outputs. The definition requires that, in the same trace, there exists a sequence $\kappa_n, \ldots \kappa_0$ where each π_i and κ_i are complementary in that one is an input, the other is an output, and the input and output messages match. We also insist the channel on which each κ_i performs and input or output is equivalent to $f(K)$ according to the equational theory.

Notice we make use of $\text{State}_B \models u_i = M_m$ to state that these messages match since State_B knows all the outputs in its extended process. This allows the attacker to construct the message in a different way other than providing directly u_i. For example, if u_i maps to a constant *hello* the attacker can directly provide the input message *hello*, without violating agreement. Furthermore, since the message *hello* was already known to the attacker they may induce the input before the output is sent in the trace, also without violating agreement. That is, there is no requirement that outputs occur before their corresponding inputs for the agreement property.

5.2.1 Example attack on agreement

We now consider some examples of attacks on agreement. Consider the protocol Agreement-Example, depicted in the MSC in Fig. 5.1. As the chart suggests, this protocol contains an invalid agreement claim. The protocol also makes use of digital signatures, which were touched on previously in Sec. 2.3.

We start by extending the syntax of our message terms to include some necessary terms for the notions of signing messages and checking signatures.

5.2 Agreement – a strong authentication property

$$\begin{aligned}
\text{M, N} ::=\ & x & \text{variable} \\
 & |\ \text{pk(M)} & \text{public key} \\
 & |\ \{\text{M}\}_\text{N} & \text{encryption} \\
 & |\ \text{dec(M, N)} & \text{decryption} \\
 & |\ (\text{M, N}) & \text{tuple} \\
 & |\ \text{fst(M)} & \text{first projection} \\
 & |\ \text{snd(M)} & \text{second projection} \\
 & |\ \text{sign(M, N)} & \text{signature} \\
 & |\ \text{check(M, N)} & \text{signature check} \\
 & |\ \text{h(M)} & \text{hash}
\end{aligned}$$

(EXTENDED SYNTAX OF MESSAGES)

We model signatures such that signatures are formed using a message and a private key, which typically will be unknown to the adversary when the signing agent is honest. Signed messages are checked by checking that the public key returns the intended signed message, as modelled by extending E with the following equation.

$$\text{check}(\text{sign}(M, K), \text{pk}(K)) =_\text{E} M \qquad \text{(CHECK)}$$

Notice that there is no destructor for cryptographic hash functions. This models the assumption in the Dolev-Yao model that cryptographic hashes cannot be reversed in reasonable time.

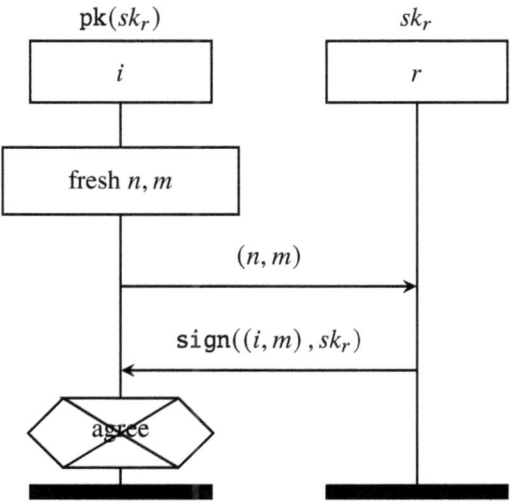

Fig. 5.1 Agreement-Example protocol.

The roles of the above protocol can be specified as follows. You can see that the initiator checks the signature against the expected message using the appropriate public key.

$$Initiator_A(c_i, i, r, pk_r) \triangleq \mathsf{new}\ n, m;\\ \mathsf{out}(c_i, (n, m));\\ \mathsf{in}(c_i, x);\\ \mathsf{if}\ \mathsf{check}(x, pk_r) = (i, m)\ \mathsf{then}\\ \mathsf{agree}(c_i); 0$$

$$Responder_A(c_r, r, sk_r, i) \triangleq \mathsf{in}(c_r, y);\\ \mathsf{out}(c_r, \mathsf{sign}((i, \mathsf{snd}(y)), sk_r)); 0$$

We introduce an initial configuration where each session is on a fresh channel. This will allow us to distinguish executions of one session from another. Initially, for simplicity, we provide an "enterprise architecture" threat model, but will introduce an "open network" threat model for later examples.

new sk_b;
out($keys$, pk(sk_b)); (
(! new c_i; out($init, c_i$);
   $Initiator_A(c_i, a, b, \mathsf{pk}(sk_b))$) | (EXTENDED MULTI-CHANNEL CONFIGURATION)
(! new c_r; out($resp, c_r$);
   $Responder_A(c_r, b, sk_b, a)$))

After two transitions we reach the following initial state from which we will analyse attacks.

$$\mathsf{new}\ sk_b; \left[\{ pk_b \mapsto \mathsf{pk}(sk_b) \} \left| \begin{array}{l} (!\,\mathsf{new}\ c_i; \mathsf{out}(init, c_i);\\ \quad Initiator_A(c_i, a, b, \mathsf{pk}(sk_b))) \mid\\ (!\,\mathsf{new}\ c_r; \mathsf{out}(resp, c_r);\\ \quad Responder_A(c_r, b, sk_b, a)) \end{array} \right. \right]$$

(STATE MULTI-CHANNEL-2)

In order to enable the derivation of traces containing agreement and synchronisation claims, we extend our operational semantics (cf. Section 3.3.3) with the following rule.

$$\frac{M =_E N}{\mathsf{agree}(M); P \xrightarrow{\mathsf{agree}(N)} [id \mid P]}\ (\text{AGREE})$$

Based on the previous definition, we now demonstrate an attack on the agreement claim of the protocol Agreement-Example, which is shown informally as an MSC in Fig. 5.2 and formally as a trace below.

5.2 Agreement – a strong authentication property

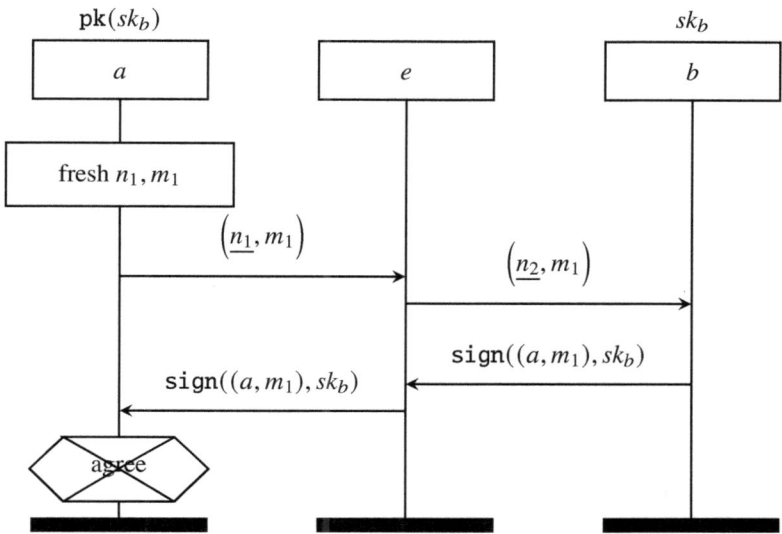

Fig. 5.2 An attack on agreement for Agreement-Example, MSC representation.

$$\text{new } sk_b; \left[\frac{\{pk_b \mapsto \text{pk}(sk_b)\}}{\begin{array}{c} (!\,\text{new } c_i; \text{out}(init, c_i); \\ Initiator_A(c_i, a, b, \text{pk}(sk_b))) \mid \\ (!\,\text{new } c_r; \text{out}(resp, c_r); \\ Responder_A(c_r, b, sk_b, a)) \end{array}} \right] \models \begin{array}{l} \langle \text{out}(init, c_i) \rangle \\ \langle \text{out}(c_i, u_1) \rangle \\ \langle \text{out}(resp, c_r) \rangle \\ \langle \text{in}(c_r, (n_2, \text{snd}(u_1))) \rangle \\ \langle \text{out}(c_r, u_2) \rangle \\ \langle \text{in}(c_i, u_2) \rangle \\ \langle \text{agree}(c_i) \rangle \\ ((n_2, \text{snd}(u_1)) \neq u_1) \end{array}$$

In this attack, the adversary intercepts the initial message from a to b, and changes it by changing n_1 and replacing with it with n_2, as underscored in Fig. 5.2. The attacker then simply relays back the message from b to a. Since the message from b does not depend on the value n_1, a believes that she and b agree on the value of n, which is not true: a believes that b received n_1, whereas b believes that a sent n_2. To emphasise where the disagreement on message is in the trace above, we indicate the inequality $(n_2, \text{snd}(u_1)) \neq u_1$ at the end of the formula above.

Mobility. Notice in the threat model introduced to model agreement that channels named *init* and *resp* are used to output fresh names that are also used as channels. For a deeper understanding, pay attention to how extruded names are handled when using the rules of the operational semantics in Fig. 3.6 to derive the following labelled transitions.

$$[id \mid \text{new } c; \text{out}(init, c); \text{new } m; \text{out}(c, m)]$$
$$\xrightarrow{\text{out}(init,ch)} \text{new } c; [\{ch \mapsto c\} \mid \text{new } m; \text{out}(c, m)]$$
$$\xrightarrow{\text{out}(ch,u)} \text{new } c, m; [\{ch \mapsto c, u \mapsto m\} \mid 0]$$

Notice that the same mechanisms used to output nonce m are also used to output the channel c. Like the nonce, the alias on the output label is different from the name binding the channel. By sending a fresh channel in this way, the published channel becomes available to attackers. In contrast, process new c, m; out(c, m) cannot make any transition, since fresh channel c has not yet been published. This is an example of a sophisticated feature of π-calculi, called mobility, where channels may be passed around like data [91, 92].

5.2.2 Agreement in ProVerif

ProVerif and related tools can leverage queries involving an implication to model agreement, as we will explain here. ProVerif has injectivity checks built in and hence perhaps encoding injective agreement in ProVerif is easier than the more direct definition we gave (cf. Def. 5.3). The direct definition remains useful, since it can guide you towards writing down the correct query in ProVerif in general.

As usual we declare the equational theory. For this example, an equational theory for signatures suffices.

type pubkey.
type seckey.
fun pk(seckey): pubkey.
fun sign(bitstring, seckey): bitstring.
reduc forall x: bitstring, y: seckey; check(sign(x, y), pk(y)) = x.

We now declare events and a query that will check for agreement when the events are positioned appropriately in the code.

event agree(bitstring).
event sentLastMessage(bitstring).
query ms: bitstring;
   inj-event(agree(ms)) $\implies$ **inj-event**(sentLastMessage(ms)).

In the above, there is an implication saying that if we have agreed on the messages ms then the communicating partner has sent their last message prior to that and agrees on the same messages. The prefix **inj-** can be removed to obtain agreement in its non-injective form. This form of query is called a *correspondence assertion* [34].

5.2 Agreement – a strong authentication property

The events above are then used to annotate the roles in the protocol. In the initiator role below, agreement appears at the end of the thread once all checks have been performed (and if the communicating partner is assumed to be honest). The agree event lists the two messages sent and received by that role.

let *Initiator*(*ishonest*: bool, c: channel, *i*: bitstring, *r*: bitstring, *pkr*: pubkey) =
　new *n1*: bitstring;
　new *n2*: bitstring;
　let *m1*: bitstring = (*n1*, *n2*) **in**
　out(c, (*n1*, *n2*));
　in(c, *m2*: bitstring);
　if check(*m2*, *pkr*) = (*i*, *n2*) **then**
　if *ishonest* **then**
　event agree((*m1*, *m2*)).

The counterpart role by the responder is annotated by an event positioned such that it must have occurred if the responder has sent its last message to the initiator. The messages received and sent are parameters of the event.

let *Responder*(c: channel, *r*: bitstring, *i*: bitstring, *skr*: seckey) =
　in(c, *m1*: bitstring);
　let (*n1*: bitstring, *n2*: bitstring) = *m1* **in**
　let *m2* = sign((*i*, *n2*), *skr*) **in**
　event sentLastMessage((*m1*, *m2*));
　out(c, *m2*).

Notice that the query insists that the list of messages are the same, and hence the same inputs and outputs appears in the same order on both sides. This captures the idea in Definition 5.3, that the pair of messages sent and received should be the same. Furthermore, by gathering together all messages exchanged in one session as a list of messages, we no longer require a trick to identify sessions such as creating new channels to uniquely identify each session.

For a more complete picture, reusable in examples going forward, we assume that initiator is open to connections with compromised agents. The following use of an open session should be familiar from previous sections.

let *openInitiator*(c: channel, a: bitstring, b: bitstring, *pkb*: pubkey) =
　in(c, *e*: bitstring);
　if *e* = b **then**
　　Initiator(true, c, a, b, *pkb*)
　else
　　in(c, *pke*: pubkey);
　　Initiator(false, c, a, *e*, *pke*).

free c: channel.

free a, b: bitstring.
process new *skb*: seckey;
  out(c, pk(*skb*)); (
  (!*openInitiator*(c, a, b, pk(*skb*))) |
  (!**in**(c, *e*: bitstring);
    Responder(c, a, *e*, *skb*)))

Using the code above, ProVerif generates a trace with an **agree** event listing messages exchanged and a corresponding event by the responder. One of the messages received by the responder should be different from the message sent by the initiator. You can compare the output of ProVerif to the attack vector in the MSC in Fig. 5.2 and corresponding modal logic formula presented previously.

Remark 5.1 Definition 5.3 and the formulation of agreement in ProVerif shown above are expressed differently but are not so far apart. In Def. 5.3, we emphasise that we pick out the sequence of inputs and outputs for the thread that is authenticating and the threat that is being authenticated. The use of channels helps pick out the messages that are involved in a single execution of a thread. In the ProVerif code above, we have picked out those inputs and outputs manually and listed them inside the authentication event, rather than simply listing a channel unique to the session. We required expertise to know how to set up and pick out those messages, while Def. 5.3 tells us what inputs and outputs must be considered. This is also the reason why the MSCs can simply shows a single word without listing messages. In ProVerif, we listed manually the corresponding inputs and outputs in a separate event, before the last output. ProVerif is then geared towards verifying such a property, and has the built-in features discussed above to enforce injectivity. Since both styles have their advantages, both are presented.

5.2.3 Attack on agreement requiring an open network

Now consider the (non-injective) agreement claim of the Yo protocol, depicted below in Fig. 5.3.

We first show the attack formally below. For this example, we give the attacker some power of an open network, by allowing any agent to connect to the responder (defining threads *Initiator*$_{Yo}$ and *Responder*$_{Yo}$ is left to the reader).

5.2 Agreement – a strong authentication property

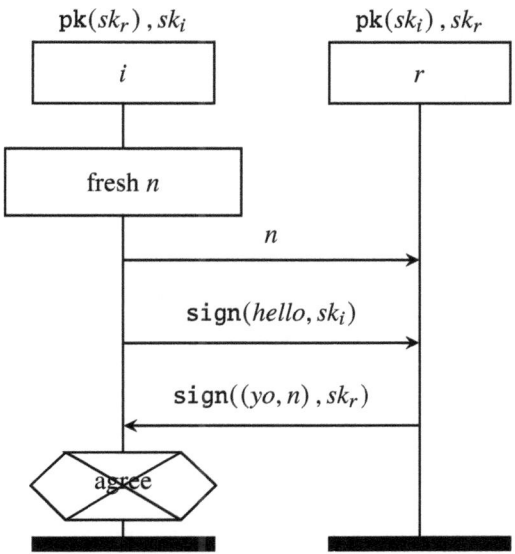

Fig. 5.3 The Yo protocol.

$$\left[\begin{array}{l} \text{new } sk_a, sk_b; \\ \left\{ \begin{array}{l} pk_a \mapsto \text{pk}(sk_a), \\ pk_b \mapsto \text{pk}(sk_b) \end{array} \right\} \\ \hline (!\,\text{new } c_i; \text{out}(init, c_i); \\ \quad Initiator_{Y_o}(c_i, a, sk_a, b, \text{pk}(sk_b))) \mid \\ (!\,\text{new } c_r; \text{out}(resp, c_r); \\ \quad \text{in}(c_r, e); \text{if } e = a \text{ then} \\ \quad\quad Responder_{Y_o}(c_r, b, sk_b, a, \text{pk}(sk_a)) \\ \quad \text{else in}(c, pk_e); \\ \quad\quad Responder_{Y_o}(c_r, b, sk_b, e, pk_e)) \end{array} \right] \models \begin{array}{l} \langle \text{out}(init, c_i) \rangle \\ \langle \text{out}(c_i, u_1) \rangle \\ \langle \text{out}(c_i, u_2) \rangle \\ \langle \text{out}(resp, c_r) \rangle \\ \langle \text{in}(c_r, e) \rangle \\ \langle \text{in}(c_r, \text{pk}(sk_e)) \rangle \\ \langle \text{in}(c_r, u_1) \rangle \\ \langle \text{in}(c_r, \text{sign}(hello, sk_e)) \rangle \\ \langle \text{out}(c_r, u_3) \rangle \\ \langle \text{in}(c_i, u_3) \rangle \\ \langle \text{agree}(c_i) \rangle \\ (\text{sign}(hello, sk_e) \neq u_2) \end{array}$$

In the above trace, we see that the problem emerges from the fact that although an exchange of messages occurs in the appropriate order (as depicted in the MSC in Fig. 5.4), b never knows that a tried to run the protocol with him. Instead, b believes only e ran the protocol with him. Thus, an adversary can relay the messages from b intended for e to a instead, thereby tricking a into completing the protocol. She believes that she was in a direct conversation with b, but in fact b was talking to e. The problem is reflected by the fact that, for this trace and any function $f: c_i \mapsto c_r$, the second message sent by a, named u_2 in the trace above, does not match the

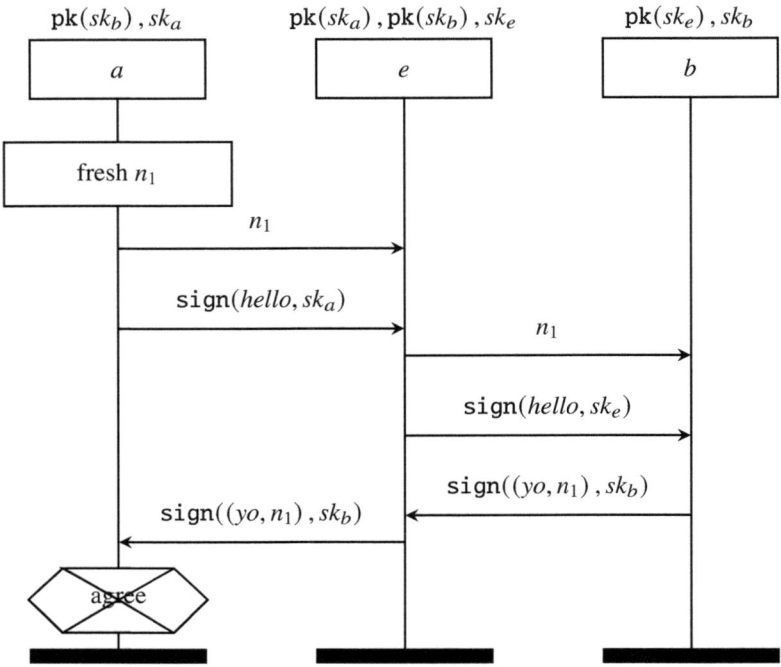

Fig. 5.4 Attack on non-injective synchronisation for the Yo protocol.

second message received by b, which was constructed directly as $\mathsf{sign}(\mathit{hello}, \mathit{sk_e})$. We indicated this as a hint at the end of the trace.

5.2.4 Attack on injectivity

Protocols satisfying agreement, but not their injective variants, are susceptible to replay attacks. In these attacks, the adversary waits for a correct execution between two honest agents to occur, and then uses the elicited information to authenticate with an honest agent in another run of the protocol. The notion of injective agreement addresses this problem. An example of such an attack was discussed informally previously in Fig. 2.1.

We demonstrate an attack on injective agreement on another unilateral authentication protocol, depicted by the MSC in Fig. 5.5.

The initiator and responder are defined as follows. Notice the responder is making the agreement claim.

5.2 Agreement – a strong authentication property

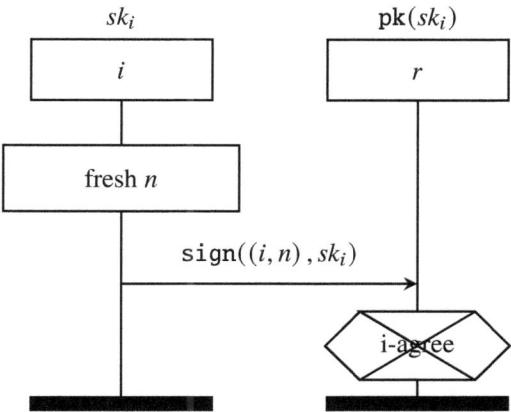

Fig. 5.5 Unilateral authentication.

$$Initiator_U(c_i, i, sk_i, r) \triangleq \text{new } n;$$
$$\text{out}(c_i, \text{sign}((i, n), sk_i)); 0$$

$$Responder_U(c_r, r, i, pk_i) \triangleq \text{in}(c_r, y);$$
$$\text{if } \text{fst}(\text{check}(y, pk_i)) = i \text{ then}$$
$$\text{agree}(i, c_i); 0$$

The configuration and replay attack are described formally below.

$$\text{new } sk_a; \begin{bmatrix} \{pk_a \mapsto \text{pk}(sk_a)\} \\ \hline (!\,\text{new } c_i; \text{out}(init, c_i); \\ Initiator_U(c_i, a, sk_a, b)) \mid \\ (!\,\text{new } c_r; \text{out}(resp, c_r); \\ Responder_U(c_r, b, a, \text{pk}(sk_a))) \end{bmatrix} \models \begin{array}{l} \langle \text{out}(init, c_i) \rangle \\ \langle \text{out}(c_i, u_1) \rangle \\ \langle \text{out}(resp, c_r^1) \rangle \\ \langle \text{in}(c_r^1, u_1) \rangle \\ \langle \text{agree}(c_r^1) \rangle \\ \langle \text{out}(resp, c_r^2) \rangle \\ \langle \text{in}(c_r^2, u_1) \rangle \\ \langle \text{agree}(c_r^1) \rangle \\ \text{true} \end{array}$$

In the above attack, firstly a correctly authenticates b, but then the adversary replays the same message to b, making him believe that he authenticated a again, while a did not attempt a second run. Observe that by the time we reach the end of the trace any function f mapping sessions of a to sessions on an honest agent being authenticated (b) is such that $f: c_r^1 \mapsto c_i$ and $f: c_r^2 \mapsto c_i$, which is not injective, and such mappings are the only possibility for satisfying the conditions of agreement.

5.2.5 Practice in verifying agreement

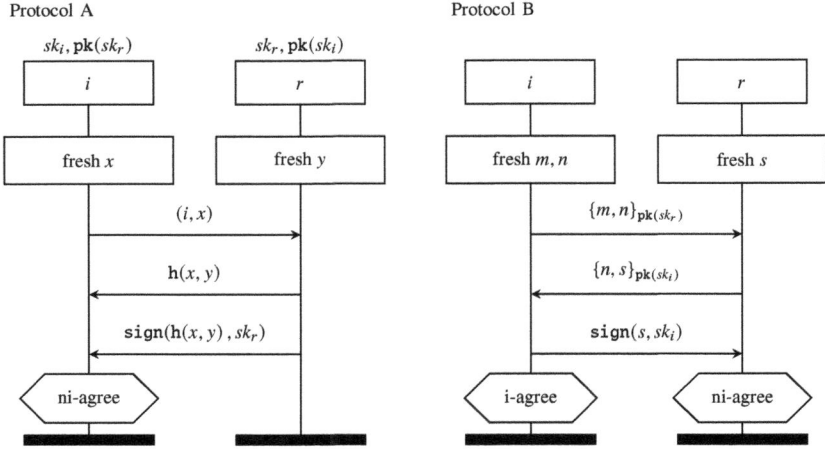

Fig. 5.6 Protocols for the exercises on agreement.

Exercise 5.1 Answer the following questions as fully as possible, providing formal specifications and attacks where possible.

1. Consider the protocol A in Fig. 5.6. Provide an attack for the ni-agree claim.
2. Consider the protocol B in Fig. 5.6. Provide an attack for the ni-agree claim from the perspective of the responder. Can you find an attack for the i-agree claim for the initiator? Explain your answer as precisely as possible.
3. Model the above examples in ProVerif and use ProVerif to check your examples. Make use of the example ProVerif code by modifying it to change the protocol and ensure that the message listed in the events reflect all the messages sent and received. Recall that you can toggle between injective and non-injective agreement using the prefix `inj-` in the query. Of course, you can switch perspectives in the protocol by positioning the events appropriately and perhaps renaming them, so that they do not clash.

5.2.6 Open networks and symmetric key cryptography

Recall now the ProVerif examples in Sec. 2.2, which provided intuition for secrecy and authentication properties of some variants of the BAC protocol used in ePassports. Now would be an appropriate moment to revisit the examples in that section with the more solid grounding obtained thus far.

5.2 Agreement – a strong authentication property

Elsewhere we have employed public key cryptography, primarily because we aimed to simplify matters by avoiding confusion caused by switching between different message theories for encryption. There was also a deeper reason for looking first at symmetric cryptography in Sec. 2.2 before diving deeper into public key cryptography from Chapter 3. Specifically, the open network model is typically redundant if only symmetric keys are used in a protocol. To see why, recall that in the open network threat model we assume that an honest agent can interact with both honest and malicious agents. In the symmetric key setting, this means that an honest agent a can run a thread with a both a shared secret with another honest agent b, and also a shared secret with a dishonest agent e. The attacker does not know shared secrets between a and b and hence the sessions between A and B need to be modelled. However, the attacker does know the secrets shared between a and e. In particular, this means that the attacker can perform all actions that a would perform when interacting with e and therefore that process need not be modelled explicitly in the network. This simplifies protocol analysis in the purely symmetric key setting.

For example, consider the following where honest a runs A^{honest} containing a secrecy or authentication claim at the end of each thread, and honest agent b runs B with the same keys.

$$\text{new } k; ((!\text{new } n; A^{honest}(k,n)) \mid (!\text{new } m; B(k,m))) \mid$$
$$(!\text{new } n; A^{compromised}(k',n))$$

In the above, k is the shared secret and n and m are the nonces generated fresh for each session. Also, below, observe that there is a process $A^{compromised}$ that models the interaction between honest agent a and some dishonest agent e. The fact that a interacts with a dishonest agent is modelled by letting the shared secret k' be a free variable, and also omitting any secrecy or authentication claim (as mentioned above, claims in honest threads are inserted as normal).

Process $!\text{new } n; A^{compromised}(k',n)$, above, can be safely removed, since the effect of each input or output stemming from that network component on any trace can be accounted for in other actions. Specifically, every time an output is used that originates from the given component is used in a trace then the effect of using that output can be simulated by a message representing how the attacker would have created that output using knowledge on the network. Since there are no security assertions in $A^{compromised}$, no observation critical for analysing secrecy or authentication will be omitted from a trace in this manner. Therefore, it is sufficient to evaluate the following network.

$$\text{new } k; \Big((!\text{new } n; A^{honest}(k,n)) \mid (!\text{new } m; B(k,m))\Big)$$

The above is indeed the form of the threat model in Sec. 2.2. The ProVerif examples in Sec. 2.2 make use of an equational theory for symmetric encryption, as well as the threat model as explained above. Thus those protocols can be formally analysed using an "enterprise" threat model, without missing anything that an "open" threat model would bring.

The above observations explain why, when using symmetric key cryptography, open sessions are not generally modelled explicitly in the network configuration. In summary, an honest agent sharing a symmetric key between itself and an attacker can be impersonated by the attacker, and hence no new information can be learnt. This ability to simplify threat models is a form of compositional reasoning, this observation is akin to techniques applied in other branches of cryptography notably universal composability [39].

We are careful to state that this observation might not generalise to all settings, e.g., where there is a mix of public key and symmetric key cryptography, or where there are multiple shared secrets distributed in various ways across multiple agents. An advanced threat model, where long-term symmetric and public keys are mixed, will be explained in Sec. 5.3.

5.3 Multiparty case study: OpenID Connect

This section guides the reader towards analysing a larger protocol involving more than two roles. The protocol also mixes public keys and symmetric keys. Protocols may involve more than one party for example if there is a party that vouches for the identity of a user who then aims to connect to another service. The methods we have presented so far can be extended to multiparty protocols. We explain here a case study involving three distinct roles for agents and draw attention to how the threat model can be set up to reflect the fact that there exist sessions of a protocol where some of the agents may be compromised.

We consider Single-Sign-On (SSO) protocols that enable users to log in to a website using another service that vouches for their identity. For example, to log in to an online store you may use your credentials stored in your Google or Facebook account, making use of standard protocols such as OpenID Connect. Companies are also trending toward using SSO protocols, for example allowing an account managed by Microsoft to log in to a range of services across an organisation. It is therefore essential that SSO protocols are secure and hence the community has developed a range of RFCs (Requests for Comments) that communicate guidelines for how to address various challenges, including the mitigation of threats.

Using the methodology of this book, it is possible to determine whether an RFC offers an effective mitigation strategy for viable threats. Threats posed to SSO protocols include Cross-Site Request Forgery (CSRF) attacks. In a CSRF attack, a user starts to authenticate with one server, while in parallel the attacker starts a session with another server. The attacker then attempts to hijack the session by giving the impression that a message from one session in fact originates in the other session. In this section, we will use tool support to expose an attack vector where the issuer gets identity providers mixed up and we also verify a known patch for that problem [58].

5.3.1 Authorization Code Grant protocol

There are many different variants of SSO protocols, sometimes called "flows" in the specifications. We consider here part of an "Authorization Code Grant" protocol, which takes care of the secure delivery of an access token to an app that allows the app to access a resource in the name of a user that approves the request. To be specific, the Authorization Code Grant protocol we analyse here is inspired by the flow provided in the Solid OIDC Primer [93], depicted in the Fig. 5.7. This is just a small part of Solid,[1] which aims to standardise protocols and policies for managing personal data online, by standardising how apps, storage, identity providers, and browsers interact [55].

There are three roles for agents in this protocol:

- **App** - The App could be a mobile app or website requesting the permission of the resource owner to access private information.
- **Browser** - The Browser is the interface of the resource owner which is used to connect both the App and Issuer, allowing them to talk via HTTPS requests. The resource owner knows the password required to testify their identity, modelled by the Browser having the password available to them when required.
- **Issuer** - The Issuer vouches for the identity of the resource provider, which should avoid the password from being transmitted to the App.

After having declared the roles, our model skips the first few steps of the primer, which setup communication between the agents. The 'Fetch Client ID Document' step is modelled by assuming that App has already looked up the user and has relevant public information about honest agents in their initial knowledge, as we will model using a suitable open network model.

Since HTTPS communications are used heavily in this protocol, some nonstandard notation is used to model HTTPS handshakes. Before starting communication and sending any message, the two actors in question are always using an HTTPS handshake to establish a connection. These handshakes, defined by TLS, guarantee only unilateral authentication. In particular, the client authenticates the server but not the other way round (hence the need to run an authentication protocol on top of TLS). The little arrow in the MSC in Fig. 5.7 indicates which party is actually authenticated by running the handshake. Using the HTTPS handshake before every communication between two agents also mimics the use of callbacks in the primer. A callback in Web apps is an HTTPS URI that is used to respond to a request, e.g., for submitting data from a Web form during a redirect.

The handshake in HTTPS is conducted using TLS which makes use of the certificate of the domain. Thus, when handshakes rely on HTTPS we should be aware that we are relying on the assumption that the entire domain is sufficiently well managed that the entire domain can be trusted with this data. There are risks associated with HTTPS such as Trojan horses and insider threats compromising the domain from within, so to rely on HTTPS, we must trust the organisation managing the domain.

---

[1] The Solid project: `https://solidproject.org/`

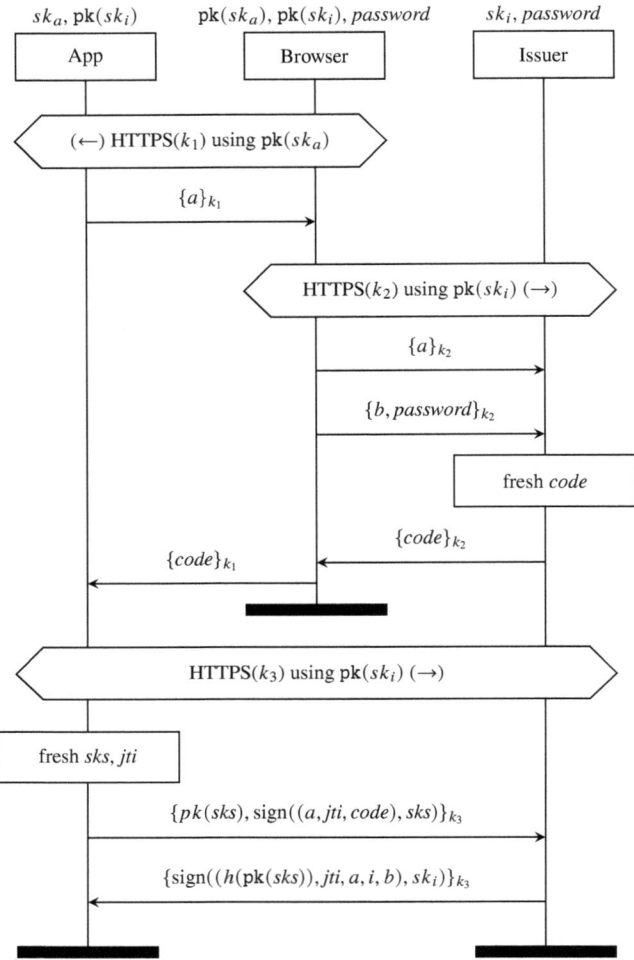

Fig. 5.7 A protocol based on the Authorisation Code Grant flow.

The first message actually being sent is the authorisation request containing the ID of the app a. There are also identities for the browser b, representing the username of the user logging in via the browser, and the name of the issuer i representing a URI where the issuer can be accessed. Some information documented for this flow is omitted, such as the scope indicating what is accessed.

The app first indicates to the browser that they would like access to be granted, at which point the browser redirects to the identity provider. The user of the browser then enters their credentials, which are checked by the issuer. In the figure, the

5.3 Multiparty case study: OpenID Connect

password is modelled as a shared secret between the browser and the issuer. If the credentials match, the issuer generates a code, which is a nonce that is sent via the browser to the app, via another redirect. The app uses the code to request a token called a DPoP token [57], which can later be used for resource access. The DPoP token comprises of a message signed by the issuer containing a public key corresponding a secret key chosen by the app. This way the app can provide "Proof of Possession" of the token at a later stage in the protocol. We stop at this point, since this part of the protocol is sufficiently interesting, as we will see in the exercises.

5.3.2 An identity provider mixup attack

We will analyse some variants of this protocol to see which are secure according to the threat models developed in this book. We will see that some are secure and others are not.

The protocol in Fig. 5.7 is known to be vulnerable to an identity provider mixup attack [58]. The key to this attack is that an honest app starts a session (with any browser) assuming that the user will log in to an identity provider that is compromised by the attacker. The attacker can then abuse their position to trick an honest user to log in to another identity provider that is not compromised and get the honest identity provider and app to erroneously provide secret information to the attacker that can be used by the attacker to log in as if they were the honest user connected to the honest identity provider.

An attack vector is presented in Fig. 5.8. Notice that a message containing the secret *code* is returned to the eavesdropper. This is because in the first step we assume that the app is connecting to some browser, and using the eavesdropper as an identity provider. The thing is that, in contrast, the honest issuer and browser both think that they are in a session where everyone is honest, and hence assumes all is well. This is the mix up resulting in the failure of a secrecy query on the side of the browser. The failure of this secrecy query is an important indicator that there is a problem, since the code is the last thing that the browser handles and the user in the browser should be confident that it remains confidential going forwards, given that the user of the browser believes that they are interacting with trusted agents.

You can see that there is also a failure of secrecy of the signed certificate (the DPoP Token) from the perspective of the issuer. This is because the attacker has the code and can make up its own session key, say sks'. The issuer then assumes that the message it receives comes from the honest app and, unwittingly, serves the signature to attacker using the attacker's own session key.[2]

---

[2] Worse still, since the attacker chose the secret key, this would cause problems later in the protocol. However, modelling the next stage of the protocol is left as an exercise for readers seeking a challenging problem.

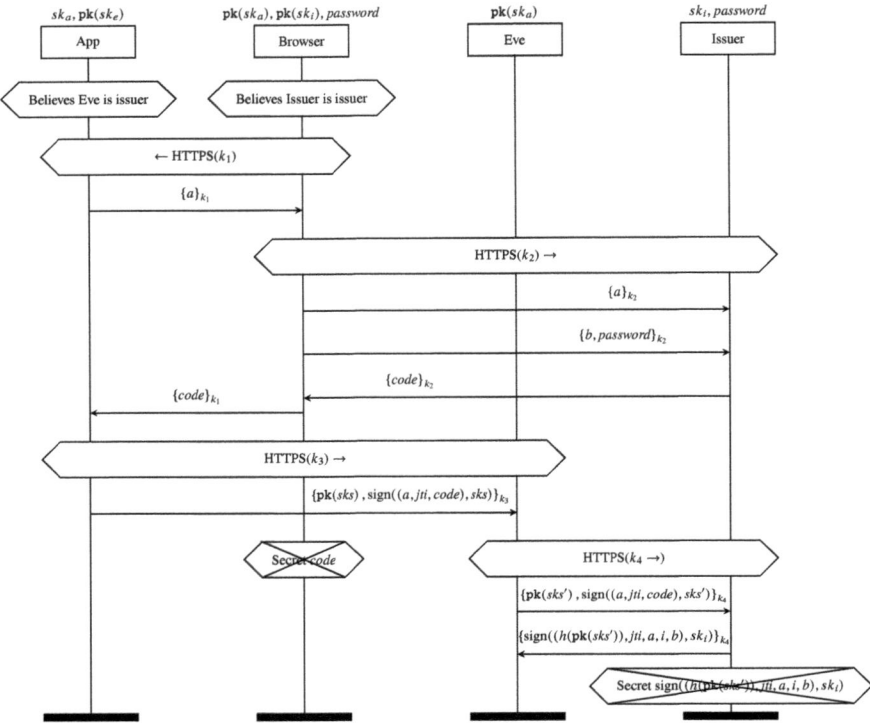

Fig. 5.8 Attack on the secrecy of the code from the perspective of the browser.

5.3.3 Practice in multiparty protocols and ProVerif

This case study is sufficiently complex that when we model it in ProVerif we require some tricks. We now walk through a series of protocol variants and threat models based on OIDC.

Code is provided here inline to help get started, reproduced also via Code Inside (https://github.com/sn-code-inside/security-protocols-and-threat-models). However, the code provided is not expected to terminate initially until you follow through with the steps directed by the exercises. We require a mix of symmetric key and public key cryptography, for which the rules you will know how to define from previous exercises. In the model of signatures below getMsg allows the message to be obtained from a signature even without the public key; therefore sign models both the message being signed and the signature itself.

type symkey.
fun senc(bitstring, symkey): bitstring.
reduc forall x: bitstring, y: symkey; sdec(senc(x, y), y) = x.

5.3 Multiparty case study: OpenID Connect

type seckey.
type pubkey.
fun pk(seckey): pubkey.
fun aenc(bitstring, pubkey): bitstring.
reduc forall x: bitstring, y: seckey; adec(aenc(x, pk(y)), y) = x.
fun hash_pk(pubkey): bitstring.
fun sign(bitstring, seckey): bitstring.
reduc forall x: bitstring, y: seckey; check(sign(x, y), pk(y)) = x.
reduc forall x: bitstring, y: seckey; getMsg(sign(x, y)) = x.

The hash function above can be applied to public keys, as you can see in the DPoP token in the last message of Fig. 5.7. As we know, hash functions have no equations, reflecting the fact that a good cryptographic hash cannot be reversed by an attacker using reasonable time and resources.

The threads modelling roles are presented in Fig. 5.9, 5.10 and 5.11 listed at the end of this chapter. In this protocol, we use a crude approximation of an HTTPS handshake (which is sufficient for our purposes). This is an approximation of HTTPS using a simple unilateral authentication protocol allowing the client to authenticate the server and establish a channel.[3] Analysing a more complete model of HTTPS requires significant computing power.

In the thread defining the role of an app, in Fig. 5.9 the app receives a connection from the browser (establishing key *k1*). Also the app connects directly to the issuer using a handshake to establish key *k3*. The thread defining the role of the issuer receives two connection, first from the browser and then from the app. We expect secrecy to fail for this thread.

The open network assumption is more involved than for two-party protocols seen previously. There is the possibility that all agents are honest, which is the scenario where we expect secrecy to hold and hence the flag determining whether secrecy is checked is set to true, as shown below.

let *openApp*(c: channel, a: bitstring, *ska*: seckey,
               b: bitstring, i: bitstring, *pki*: pubkey) =
 in(c, *e1*: bitstring);
 in(c, *e2*: bitstring);
 if (*e1* = b) ∧ (*e2* = i) **then**
   App(true, c, a, *ska*, b, i, *pki*)
 else
   in(c, *pke2*: pubkey);
   if *e2* = i **then**
     App(false, c, a, *ska*, *e1*, i, *pki*)
   else

---

[3] It also establishes a property called forward secrecy that we have not yet covered. Since a secret session key is used to transmit the shared key, then the symmetric key cannot be recovered if the long-term public key of the server is compromised in the future.

if $e1 = $ b **then**
   $App($false$, $ c$, $ a$, ska, $ b$, e2, pke2)$
else
   $App($false$, $ c$, $ a$, ska, e1, e2, pke2)$.

Notice above that there are however three scenarios where secrecy is not expected to hold, where at least one of the other agents may not be honest. As in previous threat models, it is however important to assume that an honest app will use the correct public key for any honest agent that is present, as enforced in the above.

For the browser (Fig. 5.10) and issuer (Fig. 5.11), there is also a symmetric key that is either secret or compromised depending on whether the other agent that should also hold they key is honest or not. The open issuer is modelled as follows.

let $openIssuer($c: channel, i: bitstring, ski: seckey,
                    a: bitstring, pka: pubkey, b: bitstring, $password$: bitstring$) = $
   in(c, $e1$: bitstring);
   in(c, $e2$: bitstring);
   if $(e1 = $ a$) \wedge (e2 = $ b$)$ **then**
      $Issuer($true$, $ c$, $ i$, ski, $ a$, pka, $ b$, password)$
   else
      in(c, $pke1$: pubkey);
      in(c, $wrong\_pw$: bitstring);
      if $e2 = $ b **then**
         $Issuer($false$, $ c$, $ i$, ski, e1, pke1, $ b$, password)$
      else
      if $e1 = $ a **then**
         $Issuer($false$, $ c$, $ i$, ski, $ a$, pka, e2, wrong\_pw)$
      else
         $Issuer($false$, $ c$, $ i$, ski, e1, pke1, e2, wrong\_pw)$.

The open browser follows a similar pattern to the honest issuer, where the secret password is used only if the issuer is honest.

let $openBrowser($c: channel, b: bitstring, a: bitstring, pka: pubkey,
                    i: bitstring, pki: pubkey, $password$: bitstring$) = $
   in(c, $e1$: bitstring);
   in(c, $e2$: bitstring);
   if $(e1 = $ a$) \wedge (e2 = $ i$)$ **then**
      $Browser($true$, $ c$, $ b$, $ a$, pka, $ i$, pki, password)$
   else
      in(c, $pke1$: pubkey);
      in(c, $pke2$: pubkey);
      in(c, $wrong\_pw$: bitstring);

5.3 Multiparty case study: OpenID Connect

    if $e2 = $ i **then**
        $Browser(\text{false}, \text{c}, \text{b}, e1, pke1, \text{i}, pki, password)$
    else
    if $e1 = $ a **then**
        $Browser(\text{false}, \text{c}, \text{b}, \text{a}, pka, e2, pke2, wrong\_pw)$
    else
        $Browser(\text{false}, \text{c}, \text{b}, e1, pke1, e2, pke2, wrong\_pw)$.

The main network then comprises of the three open process running in parallel. The public keys are advertised and the password is assumed to be a secret.

free c:channel.
free a, b, i : bitstring. *(* identities *)*
process
    new *ska*: seckey;
    new *ski*: seckey;
    out(c, pk(*ska*));
    out(c, pk(*ski*));
    new *password*: bitstring; *(* Browser-Issuer secret *)*
    (
        (! *openApp*(c, a, *ska*, b, i, pk(*ski*)))
  | (! *openBrowser*(c, b, a, pk(*ska*), i, pk(*ski*), *password*))
  | (! *openIssuer*(c, i, *ski*, a, pk(*ska*), b, *password*))
)

Exercise 5.2 Try these exercises to experience how to model and analyse OIDC, plus some associated RFCs, as a substantial case study.

1. Assemble the ProVerif code explained above and add appropriate secrecy queries, as in previous examples. Run the code and observe that ProVerif does not terminate (you can kill the process after the termination warning).
   Tagging messages at different stages in the protocol can sometimes help when protocol verification appears to loop [30]. Now tag the protocol by adding a constant message to each corresponding pair of input and output message in the protocol. For example, a tag can be declared as a global constant.

   const tag1: bitstring.

   The first output can then be tagged with this constant as follows.

   let $m1 = $ aenc((tag1, n, pk(*sskI*)), *pka*) **in**
   out(c, $m1$);

As expected, the tag in the corresponding first input of the browser is handled as follows.

in(c, *m1*: bitstring);
let (*tag1'*:bitstring, *n*:bitstring, *spk1*:pubkey) = adec(*m1*, *ska*) **in**
if (tag1 = *tag1'*) **then**

Run ProVerif on the tagged protocol. From which perspective(s) is there an attack reported. Are there any undetermined outcomes as a result of running the tool? Identify which honest agent is mistaken in their belief about who they are communicating with. Who does that agent believe they are communicating with and who are they actually communicating with in the attack?

2. Consider RFC 9207 [105], which is intended to counter the attack discovered above. To implement this fix, add the identity of the issuer to the two messages in Fig. 5.7 exchanging the code sent from the Issuer to the App via the Browser. Run the tool to check that attack vectors have been mitigated by this measure. In what sense is this fix is similar to the fix by Lowe for the Needham-Schroeder protocol? Check also secrecy of the whole DPoP token from the perspective of the app.
3. Consider a different countermeasure, defined by RFC 7636 [93, 102]. In this RFC, a nonce (code_challenge) is generated and hashed by the App and the hashed nonce (code_verifier = h(code_challenge)) is sent along with the first message. The Browser forwards the hashed nonce to the Issuer along with the password then the Browser contacts the Issuer. Later, in the message where the App contacts the Issuer directly in Fig. 5.7, the nonce is provided without being hashed. The Issuer can then hash that nonce to check whether the hash received earlier matches.
Model this variant of the protocol (based on the code from part 1 of this exercise). For what secrecy properties is an attack discovered, and is the attack different from any attack on the previous two models?

5.3.4 Practice in multiparty agreement

The next advanced exercise transforms our case study such that we can verify multiparty agreement. So far, we have considered only agreement between two parties. When there are more than two parties, when verifying an agreement claim if an honest partly completes the protocol with honest participants, then, ideally, all other honest participants also followed the protocol up to that point. This is trickier to express in general than two-party agreement (Def. 5.3), where there was a straightforward pairwise input/output match between both parties. Multiparty agreement we illustrate, by example, using OIDC.

The exercises in Sec. 5.3.3 already showcased examples complex enough to challenge tools. If you try to assert the secrecy of the *code* from the perspective of

5.3 Multiparty case study: OpenID Connect

the issuer any of the models developed in Exercise 5.2, part 1, ProVerif may return "cannot be proved".[4] We will show here that, however, multiparty agreement from the perspective of the issuer terminates in ProVerif.

We begin by explaining multiparty authentication from the perspective of the issuer. To model this property, we will employ the following query.

event issuerCompletes(bitstring).
event browserLast(bitstring).
event appLast(bitstring).
query $m1$: bitstring, $m2$: bitstring, $m3$: bitstring, $m4$: bitstring,
$m5$: bitstring, $m6$: bitstring, $m7$: bitstring, $m8$: bitstring,
$m9$: bitstring, $m10$: bitstring, $m11$: bitstring, $m12$: bitstring;
event(issuerCompletes(($m4, m5, m6, m7, m9, m10, m11$)))
$\Longrightarrow$
event(browserLast(($m1, m2, m3, m4, m5, m6, m7, m8$))) $\wedge$
event(appLast(($m1, m2, m3, m8, m9, m10, m11$))).

To execute this query on the examples developed in the previous section, the events must be inserted in threads in the appropriate positions. Event

$$\text{\bf event}(\text{issuerCompletes}((m4, m5, m6, m7, m9, m10, m11)))$$

is inserted next to event confidential_i($sign(M, ski)$) at the end of the thread modelling the Issuer in Fig. 5.11. Notice that messages m_4, m_5, m_6, m_7, m_9, m_{10}, m_{11} listed refer to variables representing input and output messages in the thread. Events

$$\text{\bf event}(\text{appLast}((m1, m2, m3, m8, m9, m10, m11))) \quad \text{and}$$

$$\text{\bf event}(\text{browserLast}((m1, m2, m3, m4, m5, m6, m7, m8)))$$

should be inserted in the threads for the app in Fig. 5.9 and the browser in Fig. 5.10 respectively. The events must appear just before where the last message listed in the event is sent. That is, event

$$\text{\bf event}(\text{appLast}((m1, m2, m3, m8, m9, m10, m11)))$$

appears just before the action sending m_{11}, while

$$\text{\bf event}(\text{browserLast}((m1, m2, m3, m4, m5, m6, m7, m8)))$$

appears just before sending the message m_8 in the respective figures. Positioning event as such ensures that, if the respective message is sent, then the respective event will certainly be available to the query.

The query matches up pairs of input and output throughout the execution of the protocol. In particular observe that:

---

[4] Tools are evolving, so it is possible that this outcome changes, but the exercise here stands regardless.

- Messages $m9$, $m10$, and $m11$ are common to the Issuer and App.
- Messages $m4$, $m5$, $m6$, and $m7$ are common to the Issuer and Browser.
- Messages $m1$, $m2$, $m3$, and $m8$ are common to the App and the Browser.

Thus the Issuer agrees on all messages it exchanges with all other parties. In addition, the interactions between the Issuer and the App and Browser determine other messages that are not seen directly by the Issuer, yet multiparty agreement asserts that even they went well. Observe that the messages between the App and the Browser occur before the Issuer receive the last message and hence by following the chain of causal reasoning those message must have occurred is the participants are honest. Thus agreement is simply stating that for those actions that have occurred there is an agreement between the sender and receiver.

Using the above query with the event set up as described above, ProVerif should be able now to query multiparty agreement from the perspective of the issuer. ProVerif should discover an attack! Note that, for queries to terminate, we are assuming here that tagging from Exercise 5.2, part 1, is completed correctly.

Exercise 5.3 The following exercises provide further practice for multiparty agreement.

1. Position appropriate events for verifying the following query, such that the following query checks multiparty agreement from the perspective of the App.

   event appCompletes(bitstring).
   event issuerLast(bitstring).
   query $m1$: bitstring, $m2$: bitstring, $m3$: bitstring, $m4$: bitstring,
     $m5$: bitstring, $m6$: bitstring, $m7$: bitstring, $m8$: bitstring,
     $m9$: bitstring, $m10$: bitstring, $m11$: bitstring, $m12$: bitstring;
     event(appCompletes($(m1, m2, m3, m8, m9, m10, m11, m12)$))
     $\implies$
     event(browserLast($(m1, m2, m3, m4, m5, m6, m7, m8)$))$\wedge$
     event(issuerLast($(m4, m5, m6, m7, m9, m10, m11, m12)$)).

   With events correctly positioned, does this query hold or not?

2. Do the same for the following query verifying multiparty agreement from the perspective of the Browser, ensuring that the events for the App and Issuer appear just before the messages m_3 and m_7 are sent respectively.

   event browserCompletes(bitstring).
   event issuerLastToBrowser(bitstring).
   event appLastToBrowser(bitstring).
   query $m1$: bitstring, $m2$: bitstring, $m3$: bitstring, $m4$: bitstring,
     $m5$: bitstring, $m6$: bitstring, $m7$: bitstring, $m8$: bitstring,
     $m9$: bitstring, $m10$: bitstring, $m11$: bitstring, $m12$: bitstring;
     event(browserCompletes($(m1, m2, m3, m4, m5, m6, m7)$))
     $\implies$
     event(appLastToBrowser($(m1, m2, m3)$))$\wedge$

5.3 Multiparty case study: OpenID Connect

event(issuerLastToBrowser(($m4, m5, m6, m7$))).

3. Explain why issuerLastToBrowser(($m4, m5, m6, m7$)) accounts for four messages, while issuerLast(($m4, m5, m6, m7, m9, m10, m11, m12$)) encompasses eight messages. In other words, why can the Browser not assert agreement with respect to the final messages sent between the App and the Issuer, even if the App can assert agreement with respect the messages between the Browser and the Issuer?
4. Modify the running example to send the first message between the App and Browser in the clear, i.e., unencrypted. Simply replace senc(a, $k1$) with a for the app and similarly sdec($m3, k1$) with $m3$ for the browser.
   For this variant of OIDC, what does ProVerif report for: secrecy of code from the perspective of the App (can it be proven?); and agreement from the perspective of the App (can an attack be discovered?).

The reader wishing to venture further may consider trying to model the next step of the Solid OIDC protocol [93], the Request Flow. Part of that flow is an exchange where the App aims to authenticate with an "Authorization Server". This is done by the App proving possession of the secret key corresponding a public key that was signature by the issuer in the part of the protocol already presented. The Authorization Server is a new entity in the protocol that we can assume knows the public key of the honest issuer. With RFC 9207 in place, do secrecy and agreement claims hold up for that part of the protocol?

Code for starting the exercises appears in Figs 5.9, 5.10 and 5.11, listed at the end of this chapter here. For related examples of multiparty authentication protocols, see the ProVerif code linked to in related work on verifiable credentials [36].

let *App*(*talksOnlyToHonest*: bool, c: channel, a: bitstring, *ska*: seckey, b: bitstring,
              i: bitstring, *pki*: pubkey) =

(* *Serve HTTPS connection from Browser on k1* *)
in(c, *m1*: bitstring);
let (*n*: bitstring, *spkI*: pubkey) = adec(*m1*, *ska*) **in**
new *k1*: symkey;
let *m2* = aenc((*n*, *k1*), *spkI*) **in**
out(c, *m2*);

(* *Talking to Browser* *)
let *m3* = senc(a, *k1*) **in**
out(c, *m3*);

(* *Waiting For Issuer response while it talks to Browser* *)
in(c, *m8*: bitstring);
let (*code*: bitstring) = sdec(*m8*, *k1*) **in**

(* *HTTPS connection to Issuer on k3* *)
new *n2*: bitstring;
new *sskI2*: seckey;
let *m9* = aenc((*n2*, pk(*sskI2*)), *pki*) **in**
out(c, *m9*);
in(c, *m10*: bitstring);
let (*n2'*: bitstring, *k3*: symkey) = adec(*m10*, *sskI2*) **in**
if (*n2* = *n2'*) **then**

(* *Talking to Issuer* *)
new *sks*: seckey;
new *jti*: bitstring;
let *m11* = senc((pk(*sks*), sign((a, *jti*), *sks*)), *k3*) **in**
out(c, *m11*);
in(c, *m12*: bitstring);
let (*signed_M*: bitstring) = sdec(*m12*, *k3*) **in**
let *M* = check(*signed_M*, *pki*) **in**
let (*jtk*: bitstring, *jti'*: bitstring, *a'*: bitstring, *i'*: bitstring, *b'*: bitstring) = *M* **in**
if (*jtk* = hash_pk(pk(*sks*))) **then**
if (*jti* = *jti'*) **then**
if (a = *a'*) **then**
if (i = *i'*) **then**
if (b = *b'*) **then**

if *talksOnlyToHonest* **then**
event confidential_a(*code*).

Fig. 5.9 A thread modelling the App from Fig. 5.7.

5.3 Multiparty case study: OpenID Connect

let *Browser*(*talksOnlyToHonest*: bool, c: channel, b: bitstring, a: bitstring, *pka*: pubkey,
              i: bitstring, *pki*: pubkey, *password*: bitstring) =

(* HTTPS connection to App on k1 *)
new *n*: bitstring;
new *sskI*: seckey;
let *m1* = aenc((*n*, pk(*sskI*)), *pka*) **in**
out(c, *m1*);
in(c, *m2*: bitstring);
let (*n'*: bitstring, *k1*: symkey) = adec(*m2*, *sskI*) **in**
if ($n = n'$) **then**

(* Talking to App *)
in(c, *m3*: bitstring);
let *a'*: bitstring = sdec(*m3*, *k1*) **in**
if (a = a') **then**

(* HTTPS connection to Issuer on k2 *)
new *n2*: bitstring;
new *sskI*: seckey;
let *m4* = aenc((*n2*, pk(*sskI*)), *pki*) **in**
out(c, *m4*);
in(c, *m5*: bitstring);
let (*n2'*: bitstring, *k2*: symkey) = adec(*m5*, *sskI*) **in**
if ($n2 = n2'$) **then**

(* Talking to Issuer *)
let *m6* = senc((a, b, *password*), *k2*) **in**
out(c, *m6*);
in(c, *m7*: bitstring);
let (*code*: bitstring) = sdec(*m7*, *k2*) **in**

(* Talking to App *)
let *m8* = senc((*code*), *k1*) **in**
out(c, *m8*);

if *talksOnlyToHonest* **then**
event confidential_b(*code*).

Fig. 5.10 A thread modelling the Browser from Fig. 5.7.

let *Issuer*(*talksOnlyToHonest*: bool, c: channel, i: bitstring, *ski*: seckey, a: bitstring,
              pka: pubkey, b: bitstring, *password*: bitstring) =

(* Serve HTTPS connection from Browser on k2 *)
in(c, *m4*: bitstring);
let (*n*: bitstring, *spkI*: pubkey) = adec(*m4*, *ski*) **in**
new *k2*: symkey;
let *m5* = aenc((*n*, *k2*), *spkI*) **in**
out(c, *m5*);

(* Talking to Browser *)
in(c, *m6*: bitstring);
let (*a'*: bitstring, *b'*: bitstring, *password'*: bitstring) = sdec(*m6*, *k2*) **in**
if (a = *a'*) **then**
if (b = *b'*) **then**
if (*password* = *password'*) **then**
new *code*: bitstring;
let *m7* = senc((*code*), *k2*) **in**
out(c, *m7*);

(* Waiting for Browser to transfer code to App *)

(* Serve HTTPS connection from App on k3 *)
in(c, *m9*: bitstring);
let (*n2*: bitstring, *spkI2*: pubkey) = adec(*m9*, *ski*) **in**
new *k3*: symkey;
let *m10* = aenc((*n2*, *k3*), *spkI2*) **in**
out(c, *m10*);

(* Talking to App*)
in(c, *m11*: bitstring);
let (*pksks*: pubkey, *signature*: bitstring) = sdec(*m11*, *k3*) **in**
let (*a''*: bitstring, *jti*: bitstring, *code'*: bitstring) = check(*signature*, *pksks*) **in**
if (a = *a''*) **then**
if (*code* = *code'*) **then**
let *jtk* = hash_pk(*pksks*) **in**
let *M* = (*jtk*, *jti*, a, i, b) **in**
let *m12* = senc((sign(*M*, *ski*)), *k3*) **in**
out(c, *m12*);

if *talksOnlyToHonest* **then**
event confidential_i(*code*).

Fig. 5.11 A thread modelling the Issuer from Fig. 5.7.

Chapter 6
Authentication using time and distance

Abstract Temporal properties are introduced to express authentication based on recency and distance. The difference between aliveness and recent aliveness is defined, illustrated using minimal examples. Authenticating proximity is also expressed as temporal constraints on the order of events. The distance-bounding properties Mafia fraud and distance hijacking are covered. Exercises provide practice with further examples of authentication and distance-bounding protocols. Relatively new features of tools for expressing such temporal queries are touched on.

6.1 Introduction

Multiple factors may be checked within an authentication protocol. While injective agreement, covered in Chapter 5, ensures that communication partners followed the protocol, it does not state when or where they executed the messages. In this chapter we see that, by reasoning about what actions happen before other actions, it is possible to check that communication partners executed recently and even whether they were nearby. Timing aspects are covered by the classic property "recent aliveness". The more recent distance-bounding protocols exploit the fact that the speed of light cannot be hacked, to prevent relay attacks over long distances. Relay attacks for stealing cars by relaying signals between keys and car doors are already being exploited systematically by criminals to the extent that insurance premiums are increasing for some cars known to be vulnerable [108].

6.2 A rudimentary authentication property: recent aliveness

In this section, we will explore a spectrum of authentication properties guaranteeing aliveness in some sense. In its weakest form aliveness allows an agent, say a, to claim that another agent, say b, has executed some action, which is interpreted as agent b being "alive". This is the most rudimentary authentication property; and so, if aliveness fails, there is likely a serious flaw in the protocol. Not all failures of aliveness necessarily indicate a problem, if for example there is a protocol for checking certificates generated in the past, then the certificate authority need not have been recently alive during the execution of the protocol.

We recall an example extracted from Sec. 2.3 on EMV. The Offline Static Data Authentication mode in Fig. 2.10 leads a certain mode of EMV failing the property that the card was recently alive. In Fig. 6.1, we recall the Static Data Authentication mode, along with the (offline) encrypted PIN mode (Fig. 2.13) and part of the initialisation that exchanges data from the card such as the expiry date, PAN (account number), and public key of the card. Recall that, in SDA mode, the data from the card is signed using the secret key of the bank, and the terminal has the public key of the bank.

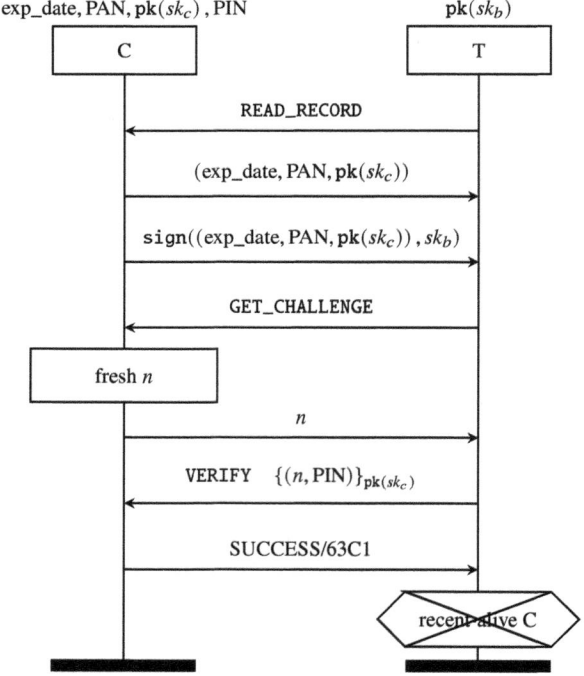

Fig. 6.1 EMV configuration featuring Static Data Authentication ODA.

SDA is a legacy mode, included in the EMV specification for backward compatibility, and should probably never be implemented. There is an attack on this mode, depicted in Fig. 6.2, where an attacker runs a session with a card, in order to learn

6.2 A rudimentary authentication property: recent aliveness

the data on the card and the signature for that data. The attacker can then replay that information at any point in order to fool the terminal into believing that the card was *recently alive*, although the conversation involving the card could have occurred a long time in the past.

This attack has very serious consequences. The consequence in the context on EMV using SDA is that it is possible to forge the card and use it until it is cancelled, solely after eavesdropping on a single session with the card. Furthermore, an honest terminal need never have been present during the initial exchange with the card in order to learn the information required to clone a card, nor need the PIN be entered. Thus an attack on recent aliveness on this protocol can be interpreted as the ability to clone cards.

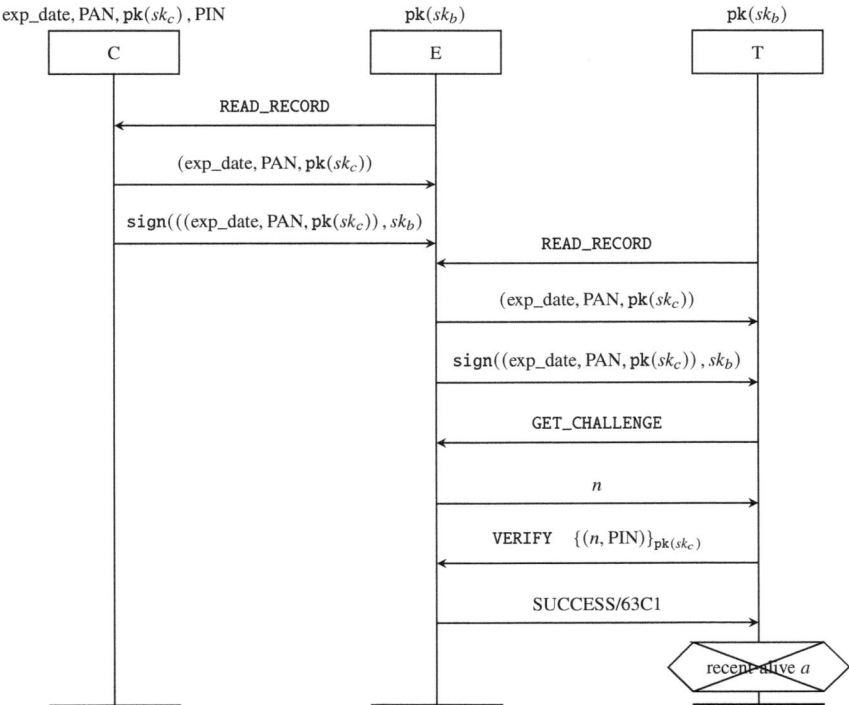

Fig. 6.2 Attack on recent aliveness of EMV with SDA.

As we explained in Sec. 2.3, there are authentication modes for the EMV protocol that mitigate such attacks on aliveness. However, if aliveness holds, it may still be that agreement fails; as we have seen for EMV cards supporting Online transactions, which we also highlighted in Sec. 2.3. Here, we focus on aliveness.

6.2.1 Assessing the extent of aliveness vulnerabilities

In this section, we consider a series of minimal protocols that fail to satisfy aliveness in different ways.

> *Variants of aliveness.* We will emphasise two flavours of aliveness.
>
> - *Weak aliveness*: makes no additional claims about agent b, other than the fact that b executed some action. In particular, weak aliveness makes no claim as to when b executed the action, nor the role b was playing when this action was executed.
> - *Recent aliveness*: allows agent a to claim that agent b executed some action during a's execution of the protocol.
>
> Additional variants of aliveness can be devised such as by insisting that an agent performs their actions in the correct role. The role can be relevant if agents may reuse their identity in different roles.

Having multiple properties is useful since, if one holds while another does not, we learn about the extent of the problem. More specifically, it is possible to have a protocol that satisfies weak aliveness, but not recent aliveness. The converse, though, it not possible: recent aliveness implies weak aliveness.

In order to formally analyse aliveness claims in our model, we need to equip ourselves with the necessary reasoning machinery, which consists of the following:

- An extension of the syntax of thread specifications to include aliveness claims, along with their operational rules.
- The definitions of validity for the new types of claims.

We first extend the syntax for processes to include aliveness claims.

P ::= out(M, N);P	send
| in(M, y);P	receive
| new x; P	fresh
| ...	etc.
| weak-alive(N); P	aliveness claim
| recent-alive(M, N); P	recent aliveness claim
	(EXTENDED SYNTAX OF PROCESSES)

Before, we could only claim secrecy. With our new syntax, we can have multiple claims at the end of threads.

In aliveness claims, the arguments are used to identify the agent. For weak aliveness it is sufficient to identify the *claimee*, that is, the agent about whom the aliveness claim is made. For recent aliveness, the first argument represents the *claimer*, that is, the agent who makes the claim, and the second one the claimee.

6.2 A rudimentary authentication property: recent aliveness

Throughout this discussion, typically, the *Initiator* claims that the *Responder* is alive or vice versa. We will use the toy protocols $Hello_0$ and $Hello_1$, depicted in the MSCs in Fig. 6.3.

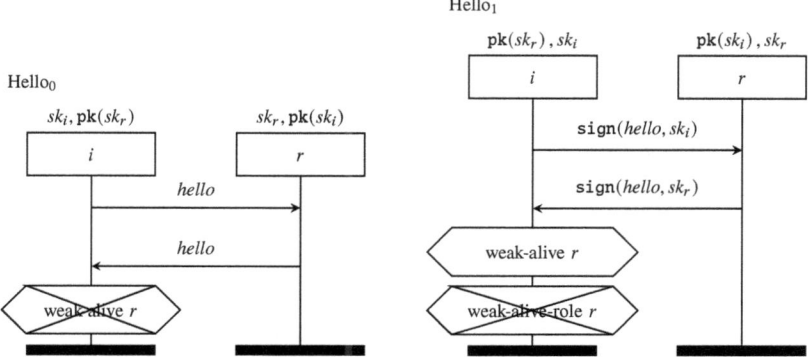

Fig. 6.3 Basic protocols failing weak aliveness and weak aliveness in the correct role.

In $Hello_0$, it is easy to see that anyone, including the adversary, can inject the "*hello*" message in the network, so receiving this message does not allow the initiator to claim that the responder is alive. The $Initiator_{H0}$ and $Responder_{H0}$ threads for $Hello_0$ are defined as follows.

$Initiator_{H0}(c, i, sk_i, r, pk_r) \triangleq$
  $out(c, hello);$
  $in(c, x);$
  if $x = hello$ then
  weak-alive$(r);$

$Responder_{H0}(c, r, sk_r, i, pk_i) \triangleq$
  $in(c, x);$
  if $x = hello$ then
  $out(c, hello);$

In analysing secrecy, we assumed a network where all messages were sent and received on a single channel, say c. For aliveness properties, similarly to agreement in Chapter 5, we will use distinct channels to differentiate actions executed by different agents. This difference compared to secrecy of a message is due to authentication properties typically asserting properties about specific actions of specific communicating partners, that, therefore, need to be identifiable.

For this simple first example, we will assume an initial configuration where a (Alice) plays the *Initiator* role and b (Bob) plays *Responder* roles. Notice in the initial configuration below, that we use the name of the responder as the channel on which the responder plays. This is so we can determine whether or not the responder acts.

```
new sk_a, sk_b;
out(keys, pk(sk_a));
out(keys, pk(sk_b));(                                    (ALIVENESS NETWORK)
(! Initiator_H0(c, a, sk_a, b, pk(sk_b)))  |
(! Responder_H0(b, b, sk_b, a, pk(sk_a))) )
```

In the above, notice the responder uses its identity as a distinct channel, making the channel distinct from global channel c. This is because the initiator wishes to know whether an action of a specific responder has occurred. The channel b allows us to pick out the actions of b.

With this initial configuration, the initial state from which we analyse this protocol is the following, where public keys have already been communicated.

$$\text{State}_{\text{Mc1}} \triangleq \text{new } sk_a, sk_b;$$
$$\left[\left\{ \begin{matrix} pk_a \mapsto \text{pk}(sk_a), \\ pk_b \mapsto \text{pk}(sk_b) \end{matrix} \right\} \middle| \begin{matrix} (!\ Initiator_{H0}(c, a, sk_a, b, \text{pk}(sk_b)))\ | \\ (!\ Responder_{H0}(b, b, sk_b, a, \text{pk}(sk_a))) \end{matrix} \right]$$

(STATE MULTI-CHANNEL-1)

We extend our operational semantics with the WEAK-ALIVE rule, defined as follows.

$$\frac{N =_E L}{\text{weak-alive}(L); P \xrightarrow{\text{weak-alive}(N)} [id\,|\,P]} \text{(WEAK-ALIVE)}$$

Unlike our definition of secrecy, executing a WEAK-ALIVE event does not unconditionally invalidate the claim. Instead, it serves to indicate that such an authentication claim is being asserted about a trace. The following definition specifies the conditions under which weak aliveness is validated.

Definition 6.1 (weak aliveness) A state State_A satisfies *weak aliveness* whenever, for all traces of the form $tr = \pi_1, \pi_2, \ldots \pi_t$, weak-alive(N) such that $\text{State}_A \xrightarrow{tr} \text{State}_B$, there exists i with $1 \leq i \leq t$ such that $\pi_i = \text{out}(L, w)$ and $\text{State}_B \models N = L$.

According to the previous definition, for a weak aliveness claim to be valid the trace must contain at least one action executed by the claimee. An execution of the claimee is identified by an output on the channel used by the claimee. With the previous definition in mind, an attack on Hello$_0$ is shown below.

$$\text{State}_{\text{Mc1}} \models \langle \text{out}(c, u) \rangle \langle \text{in}(c, hello) \rangle \langle \text{weak-alive}(b) \rangle \text{true}$$

As we can see above, the fact that the input message "*hello*" can be generated by the adversary, allows a to reach the weak aliveness claim without b having performed any action.

Such an attack on weak aliveness may be addressed by signing messages. Signing allows honest agents to ensure that the message has been created by them (thus, it could not have been created by the adversary or misinterpreted as coming from a different sender than the one expected). The Hello$_1$ protocol, in Fig. 6.3, addresses

6.2 A rudimentary authentication property: recent aliveness

problems of $Hello_0$ by making both the initiator and the responder sign the *"hello"* message with their private keys.

Indeed, if the responder role is being executed by an honest agent, sk_r is only known to that agent, so only she could have sent the second message. That is, we have that the following satisfies weak aliveness.

$$\begin{aligned}&\text{new } sk_a, sk_b; \\ &\text{out}(keys, \text{pk}(sk_a)); \\ &\text{out}(keys, \text{pk}(sk_b)); \; (\\ &\quad (!\,Initiator_{H1}(c, a, sk_a, b, \text{pk}(sk_b))) \; | \\ &\quad (!\,Responder_{H1}(b, b, sk_b, a, \text{pk}(sk_a))) \;)\end{aligned}$$

where

$$\begin{aligned}Initiator_{H1}(c, i, sk_i, r, pk_r, ch) \triangleq \; &\text{out}(c, \text{sign}(hello, sk_i)); \\ &\text{in}(c, x); \\ &\text{if } \text{check}(x, pk_r) = hello \text{ then} \\ &\text{weak-alive}(r); \, 0\end{aligned}$$

$$\begin{aligned}Responder_{H1}(c, r, sk_r, i, pk_i) \triangleq \; &\text{in}(c, x); \\ &\text{if } \text{check}(x, pk_i) = hello \text{ then} \\ &\text{out}(c, \text{sign}(hello, sk_r));; \, 0\end{aligned}$$

The above allows a to claim that b executes an action in $Hello_1$. However, if b were to may play multiple roles, it still does not allow her to claim that b is executing the responder role (weak aliveness in the correct role), as b playing the initiator role would have sent the exact same message. We can model the possibility of a and b playing multiple roles by allowing a and b to play either role. However, we only use the agent's identity to identify channels where the agent plays the role we are interested in.

$$\text{State}_{Hello_1} \triangleq$$

$$\text{new } sk_a, sk_b; \begin{bmatrix} \{pk_a \mapsto \text{pk}(sk_a), pk_b \mapsto \text{pk}(sk_b)\} \\ \hline (!\,Initiator_{H1}(c, a, sk_a, b, \text{pk}(sk_b))) \; | \\ (!\,Initiator_{H1}(c, b, sk_b, a, \text{pk}(sk_a))) \; | \\ (!\,Responder_{H1}(b, b, sk_b, a, \text{pk}(sk_a))) \; | \\ (!\,Responder_{H1}(a, a, sk_a, b, \text{pk}(sk_b))) \end{bmatrix}$$

For the above configuration, where a and b can play either role, and we insist in looking only a actions where the responder was in the correct role, there is an attack. An attack on weak aliveness of $Hello_1$ for such a configuration is shown below. Two initiators are started here.

$$\text{State}_{Hello_1} \models \begin{array}{l} \langle \text{out}(c, u_1) \rangle \\ \langle \text{out}(c, u_2) \rangle \\ \langle \text{in}(c, u_2) \rangle \\ \langle \text{weak-alive}(b) \rangle \\ \text{true} \end{array}$$

$Hello_1$'s failure to satisfy weak aliveness in the correct role is addressed in protocol $Hello_2$, shown in the MSC below, by making the contents of the messages sent in each direction different. Now, if a is executing the initiator role with b as responder, she will receive the message $\text{sign}(yo, sk_b)$, not $\text{sign}(hello, sk_b)$. If b is honest, he will not send $\text{sign}(hello, sk_b)$, and the adversary can only inject $\text{sign}(yo, sk_b)$ after it has been sent by b at least once.

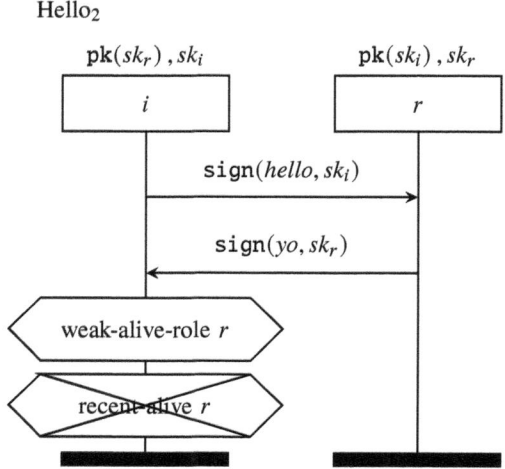

Fig. 6.4 A protocol satisfying weak aliveness in the correct roles.

However, the protocol $Hello_2$ is still not completely satisfactory from the perspective of aliveness in general. If agent a plays the initiator role, she cannot ascertain whether the agent playing the responder role was alive during execution of that session or at an arbitrarily earlier moment, i.e., $Hello_2$ fails to satisfy recent aliveness.

We will now analyse the (lack of) validity of the recent aliveness claim in $Hello_2$. The $Initiator_{H2}$ and $Responder_{H2}$ threads for $Hello_2$ are defined as follows.

$$Initiator_{H2}(c, i, sk_i, r, pk_r) \triangleq \text{ new } s; \text{out}(c, s); \\ \text{out}(s, \text{sign}(hello, sk_i)); \\ \text{in}(c, x); \\ \text{if check}(x, pk_r) = yo \text{ then} \\ \text{recent-alive}(s, r); 0$$

6.2 A rudimentary authentication property: recent aliveness

Notice above that we generate a fresh channel for each execution of the initiator, so that we are certain that we are asserting a claim about the current execution of the initiator. The fresh channel is associated with the first action of the protocol, so that we can identify when the session began when a claim is made.

$$Responder_{H2}(c, r, sk_r, i, pk_i) \triangleq \text{in}(c, x);$$
$$\text{if check}(x, pk_i) = hello \text{ then}$$
$$\text{out}(c, \text{sign}(yo, sk_r));$$

As we did for weak aliveness, we extend our operational semantics with a RECENT-ALIVE rule.

$$\frac{M =_E M' \qquad N =_E N'}{\text{recent-alive}(M, N); P \xrightarrow{\text{recent-alive}(M',N')} [id \,|\, P]} \text{ (RECENT-ALIVE)}$$

The conditions under which a recent aliveness claim is validated are given by the following definition.

Definition 6.2 (recent aliveness) A state $State_A$ satisfies recent aliveness whenever, for all traces of the form $tr = \pi_1, \pi_2 \ldots \pi_s, \text{out}(M, u), \pi_{s+1}, \ldots \pi_t, \text{recent-alive}(M, N)$ such that $State_A \xrightarrow{tr} State_B$, there exists i with $s < i \leq t$ such that $\pi_i = \text{out}(L, v)$ and $State_B \models L = N$.

According to the previous definition, for the claim to be valid, the trace must contain at least one action executed by the claimee after the claimer has executed some action in the same execution containing the claim. With the previous definition in mind, an attack on the recent aliveness claim of $Hello_2$ is shown below.

$$\text{new } sk_a, sk_b; \left[\frac{\begin{Bmatrix} pk_a \mapsto \text{pk}(sk_a), \\ pk_b \mapsto \text{pk}(sk_b) \end{Bmatrix}}{\begin{matrix}(!Initiator_{H2}(c, a, sk_a, b, \text{pk}(sk_b))) \;| \\ (!Responder_{H2}(b, a, sk_a, b, \text{pk}(sk_b))) \;|\end{matrix}} \right] \models \begin{matrix} \langle \text{out}(c, s_1) \rangle \\ \langle \text{out}(s_1, u_1) \rangle \\ \langle \text{in}(b, u_1) \rangle \\ \langle \text{out}(b, u_2) \rangle \\ \langle \text{out}(c, s_2) \rangle \\ \langle \text{out}(s_2, u_3) \rangle \\ \langle \text{in}(c, u_2) \rangle \\ \langle \text{recent-alive}(c_2, b) \rangle \\ \text{true} \end{matrix}$$

Finally, we revisit the Yo protocol in Fig. 5.3 from Sec. 5.2.3. That protocol uses a fresh nonce to uniquely distinguish each execution of the initiator role. The nonce allows the agent playing the initiator to know if a message comes from an agent playing the responder role during her execution. Yo inherits from $Hello_2$ the ability to differentiate roles, so it satisfies recent aliveness in the correct role.

In Yo, the secrecy of the nonce sent by the initiator is not guaranteed. However, this is not a problem if the protocol only intends to check the presence of the responder. Since a different nonce is used for every execution of the initiator role and the adversary only learns the value after it is sent, she cannot inject it in advance and thus the agent playing the responder role really has to be alive at that moment to be able to respond. Of course, however, the responder cannot guarantee that the initiator was recently alive.

6.2.2 Practice in verifying weak aliveness

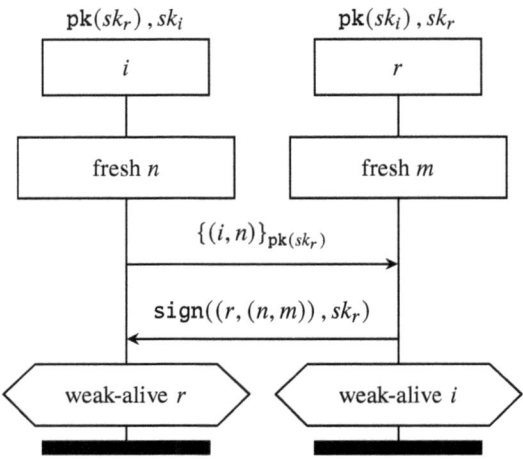

Fig. 6.5 Protocol for aliveness (Exercise 6.1).

Exercise 6.1 Consider the protocol defined by the MSC in Fig. 6.5.

1. One of the aliveness claims of the protocol holds, whereas the other does not. Indicate which one is each.
2. Formally give a trace describing an attack for the claim that does not hold.
3. Informally explain why the other claim holds.
4. Propose a fix for the claim that does not hold and informally explain why it works. Does your fixed protocol continue to satisfy the originally valid claim? Briefly explain.

6.2 A rudimentary authentication property: recent aliveness

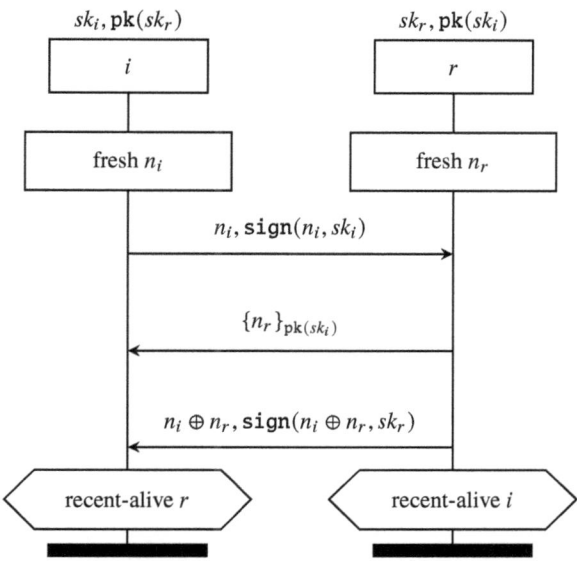

Fig. 6.6 Protocol involving XOR.

6.2.3 Practice on recent aliveness involving XOR

Consider the protocol in Fig. 6.6. This protocol involves ⊕, which is the exclusive OR (XOR) operator and which we have already encountered in Sec. 2.2.8. To make this precise, we first need to add the XOR operator to our theory of messages, as well as the constant 0, which is the unit of XOR.

$$\begin{aligned}
\mathsf{M, N} ::= &\ \ldots \\
&|\ 0 &&\text{unit of the XOR operation} \\
&|\ \mathsf{M} \oplus \mathsf{N} &&\text{XOR operation}
\end{aligned}$$

(SYNTAX OF MESSAGES EXTENDED WITH XOR)

To model this operator, extend our equational theory E to XE by including the following new equations.

$$\begin{aligned}
\mathsf{N} \oplus \mathsf{M} &=_{\mathsf{XE}} \mathsf{M} \oplus \mathsf{N} &&\text{(COMMUTATIVITY)} \\
(\mathsf{M} \oplus \mathsf{N}) \oplus K &=_{\mathsf{XE}} \mathsf{M} \oplus (\mathsf{N} \oplus K) &&\text{(ASSOCIATIVITY)} \\
\mathsf{M} \oplus \mathsf{M} &=_{\mathsf{XE}} 0 &&\text{(CANCELLATION)} \\
\mathsf{M} \oplus 0 &=_{\mathsf{XE}} \mathsf{M} &&\text{(UNITY)}
\end{aligned}$$

The equations above state that XOR is associative and commutative with identity 0 (that is, XOR is a commutative monoid), and taking a message and applying XOR with itself cancels out the message. For example, we have the following equalities by applying commutativity, associativity and cancellation.

$$(y \oplus x) \oplus y =_{\mathsf{XE}} (x \oplus y) \oplus y =_{\mathsf{XE}} x \oplus (y \oplus y) =_{\mathsf{XE}} x \oplus 0 =_{\mathsf{XE}} x$$

Consider again the protocol above involving XOR and two aliveness claims.

Exercise 6.2 Model this protocol with a threat model allowing open connections with the responder. Explain whether or not there is an attack on these recent aliveness claims, and describe formally your attacks as traces with respect to a network configuration.

Advanced. We now have a methodology for formally evaluating the recent aliveness attack on the configuration of EMV in Fig. 6.1. Model the protocol, provide a network configuration suitable for evaluating recent aliveness, and describe formally the attack in Fig. 6.2 as a formula satisfied by the configuration.

6.3 Resisting relay attacks using distance-bounding protocols

For some applications, such as contactless payments and ID tags, the security of the system is related to the physical distance between the agents involved, and not only on the messages exchanged, as studied previously. For this reason, protocols that include some kind of round trip time measurement were designed, known as *distance-bounding protocols* [16].

Relay attacks are always possible for a protocol that does not have features for distance bounding. They may also be possible for protocols with a distance-bounding feature if they are poorly designed, hence we must prove that a distance-bounding mechanism fulfils its purpose.

To take an example, the EMV protocol, as implemented in contactless-payment cards to date, does not have any distance-bounding mechanism.[1] Therefore, it is currently possible to relay messages between the card and a remote terminal using intermediate devices, such as a phone in the vicinity of the cardholder and the terminal. At the time of writing the official advice of VISA is as follows.[2]

> "To make a payment, your contactless card or payment-enabled mobile/wearable device must be placed within 2 inches of the Contactless Symbol located on the checkout terminal in order for the transaction to take place (so you can't pay accidentally)."

The above statement has long been known by security experts to be misleading if we account for threats posed by attackers. Clearly payments can be made by relaying messages from a card from any distance [49]. Although, in defence of VISA, without

---
[1] Recent EMV standards do propose a Round-Trip Relay protection mechanism that is yet to be verified with a viable implementation strategy, to our best knowledge at the time of writing [99].

[2] https://usa.visa.com/pay-with-visa/contactless-payments/contactless-payments.html

6.3 Resisting relay attacks using distance-bounding protocols

recent PIN bypass attacks [94, 25] as discussed in Sec. 2.3, security researchers were only aware of how a relay attack could be used to make small payments, which would have partly mitigated the risk of large-scale fraud.

There is an increasing array of examples where the distance of agents participating in a protocol is significant. For example, protocols for electronic car keys are typically vulnerable to distance fraud, as are probably your office keys, public transport cards, ePassports, etc. The consequences can include loss of property and identity fraud. For this reason, in our wireless world, security experts should be aware of distance-bounding protocols.

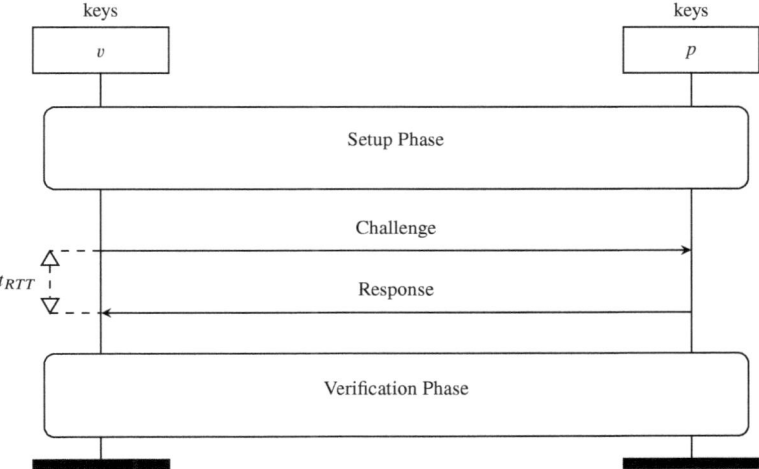

Fig. 6.7 General structure of a distance-bounding protocol.

Typically, distance-bounding protocols consist of three phases: setup phase (also called slow phase), fast phase (where the actual distance bounding occurs), and verification phase. Usually there are two roles: the Verifier and the Prover, which could be, for example, the point-of-sale terminal and the contactless payment card. The objective of the protocol is to allow the Verifier to compute a bound on the physical distance of the Prover in a cryptographically secure way. In most protocols, the Verifier sends a message (usually called challenge), and measures the time (called round trip time) until it receives a response from the Prover. Assuming processing time is negligible, the round trip time can be used to estimate the distance between the agents participating in the protocol and, as such, guarantee a *distance bound*.

Consider now the example of a distance-bounding protocol in Fig. 6.8 proposed by Meadows et al. [86]. We call this protocol Meadows 1 to distinguish it from another protocol defined by the same authors.

In this protocol, we identify and explain the three phases of the protocol below.

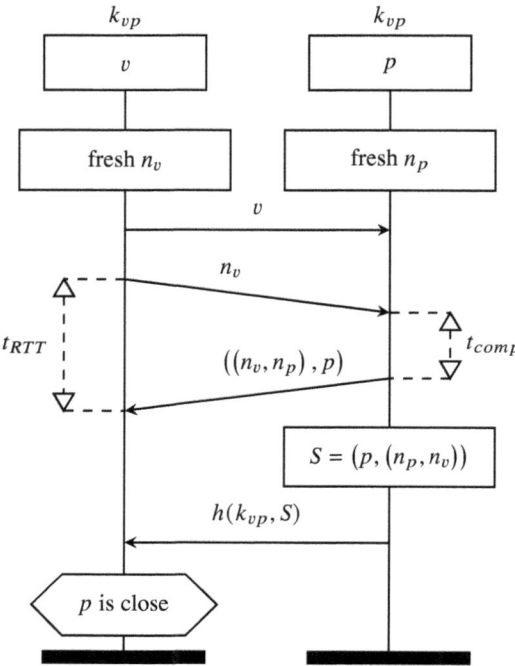

Fig. 6.8 Meadows 1 protocol.

- The *setup phase* is the first message only, which is the slow phase so there are no timing constraints. That first message sent simply says that the Verifier is ready to connect to someone, so the Prover should get ready to talk to v. In order be prepared, the Prover will generate, in advance, the nonce n_p and be ready to use it in the fast phase, since generating a cryptographically pseudo-random nonce takes non-negligible computation time. This pre-calculation of expensive cryptographic operations during the slow phase need not be explicitly modelled.
- The next two messages form the *fast phase* of the protocol. The Verifier sends out a nonce n_v, which the Prover responds to as quickly as possible by returning the nonce n_v along with n_p and the identity of the Prover p. Since the nonce n_p has already been generated, this operation can be done almost instantly upon receiving the nonce n_v. This near instantaneous response time is represented by the t_{comp}, in Fig. 6.8, which we can assume is a few nanoseconds, hence effectively zero. The interesting time t_{RTT} is the round trip time incurred by the two messages in the fast phase. In order to ensure that the Prover is near, the Verifier checks that ensure that t_{RTT} is less than the time it takes light to travel the maximum allowed distance. More precisely, the Verifier ensures the following equation holds, where

6.3 Resisting relay attacks using distance-bounding protocols

c is the speed of light and d is the maximum allowed distance.

$$c \cdot t_{RTT} \leq 2 \cdot d$$

- The final *verification phase* provides the nonces exchanged in the previous verification phases along with the identity of the Prover. A MAC (Message Authentication Code) is applied to those messages $(p, (n_p, n_v))$, which is modelled by applying a hash function h to the message along with a secret key k_{vp} which is shared by the specific agents playing the role of Verifier or Prover. Using the hash function applied to a pair, we obtain the MAC $h\left(k_{vp}, (p, (n_p, n_v))\right)$, which can only be checked by the Prover and Verifier, since the key is required to confirm the resulting hash value. This check is performed by the Verifier by checking that the second message received by the Verifier at the final step is indeed the expected hash function. If the verification phase is successful, the Verifier assumes that the Prover agrees on the nonces hence was really the agent active in the fast phase, thus is close.

The Meadows 1 protocol, described above, is generally considered to be robust against the most common problems associated with distance-bounding protocols. In the next section we make precise a key property that is enforced when we make such a claim.

6.3.1 Mafia fraud: attacks subverted by distance bounding

The setup for Mafia fraud is depicted[3] in Fig. 6.9.

> In a *Mafia fraud* the objective of an attacker is to make an honest Verifier v believe an honest Prover p is close, even though they are not. Although the honest Prover may not be close, the attacker may have devices anywhere, so can intercept messages sent by both the honest Provers and honest Verifiers and manufacture messages for them to receive.

As we have seen for authentication, there are other properties we may wish to consider. However, Mafia fraud resistance is the distance-bounding property that we must take most care of, since, if it is violated, relay attacks may be possible. Indeed, this property is violated by several distance-bounding protocols proposed in the literature, thereby those distance-bounding mechanisms can be bypassed effectively. For example, the public key version of the TREAD protocol [17] violates this property, which we will use as an illustration below.

The initial configuration for the TREAD protocol is defined as follows.

---

[3] Twemoji graphics used under the CC BY 4.0 license, without modification.

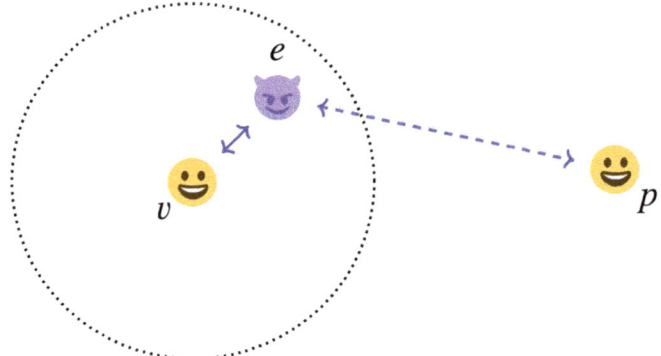

Fig. 6.9 The attacker is nearby and the honest prover is far away.

$$\begin{aligned}
\text{State}_{TREAD} \triangleq\ & \text{new } sk_a, sk_b; \\
& \text{out}(keys, \text{pk}(sk_a)); \\
& \text{out}(keys, \text{pk}(sk_b)); (\\
& (!\,\text{in}(c, e); \\
& \quad \text{if } e = b \text{ then } Prover(a, a, sk_a, b, \text{pk}(sk_b)) \\
& \quad \text{else in}(c, pk_e);\, Prover(a, a, sk_a, e, pk_e))\ | \\
& (!\,Verifier(c, b, sk_b, a, \text{pk}(sk_a)))
\end{aligned}$$

(INITIAL NETWORK CONFIGURATION)

The network contains two trusted agents a as Prover and b as Verifier. First their private keys are generated, and then the corresponding public keys are sent to the network. For the Prover, we allow it to start sessions with an arbitrary agent i. On the other hand, we only allow the Verifier b to start sessions with Prover a. This corresponds to the minimal configuration in which a Mafia fraud attack for this protocol is possible.

Before analysing our first formal example, we introduce the claim associated with the distance-bounding property.

$$\begin{aligned}
\mathsf{P} ::=\ & \ldots \\
\mid\ & \text{close}(\mathsf{M}); \mathsf{P} \qquad \text{distance bound claim}
\end{aligned}$$

(EXTENDED SYNTAX OF CLAIMS)

We extend our operational semantics with the CLOSE rule, defined as follows.

$$\dfrac{M =_\mathsf{E} N \qquad K =_\mathsf{E} L}{\text{close}(N, L); \mathsf{C} \xrightarrow{\text{close}(M, K)} [id\,|\,\mathsf{C}]} \text{ (CLOSE)}$$

The TREAD protocol is shown in Fig. 6.10. We define the threads that correspond to the *Prover* and the *Verifier* below.

6.3 Resisting relay attacks using distance-bounding protocols

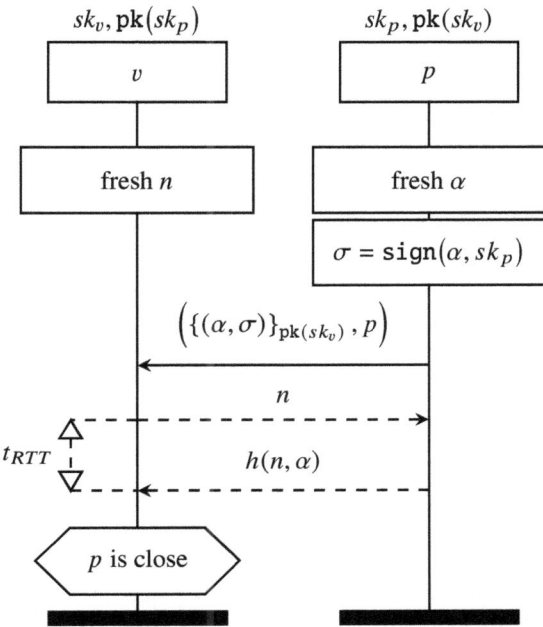

Fig. 6.10 TREAD Protocol (simplified, public key version).

$$\begin{aligned}
Prover(c, p, sk_p, v, pk_v) &\triangleq \\
&\text{new } \alpha; \\
&\text{let } \sigma = \text{sign}(\alpha, sk_p) \text{ in} \\
&\text{out}\left(c, \left(\{(\alpha, \sigma)\}_{pk_v}, p\right)\right); \\
&\text{in}(c, n); \\
&\text{out}(c, h(n, \alpha)); 0
\end{aligned}$$

When we instantiate the Prover in the network configuration, previously, notice that we used the identity of the agent playing the Prover role also as the channel. This allows actions by that agent to be picked out, much like for aliveness properties.

$Verifier(c, v, sk_v, p, pk_p) \triangleq$
    new f;　　　　　　　　　　　　　　　Channel used during fast phase.
    out(c, f);　　　　　　　　　　　　　　Make it available to attacker.
    new n;
    in(c, x);
    let $\alpha = \mathtt{fst}(\mathtt{dec}(\mathtt{fst}(x), sk_v))$ in
    let $\sigma = \mathtt{snd}(\mathtt{dec}(\mathtt{fst}(x), sk_v))$ in
    if $\mathtt{snd}(x) = p$ then
    if $\mathtt{check}(\sigma, pk_p) = \alpha$ then
    out(f, n);　　　　　　　　　　　　　　Start fast phase.
    in(f, y);　　　　　　　　　　　　　　End the fast phase.
    if $y = h(n, \alpha)$ then
    close(p, f)　　　　　　　　　　　　　Claim that agent p was close.

Notice, above, that the *Verifier* defines a channel with the sole purpose of marking the start and the end of the fast phase. This will help us when stating the security definition, next, since they will allow us to refer to the beginning and end of the fast phase in a trace.

Now we are ready to define a distance-bounding security property in terms of the order of actions. This definition is based on the distance-bounding property defined in related work [85], valid for the specific case of Mafia fraud. Executing an event close does not unconditionally invalidate the claim. The following definition specifies the conditions under which the claim is falsified.

Definition 6.3 (Mafia-fraud attack) A state State_A is vulnerable to Mafia-fraud whenever for some $tr = \pi_1, \pi_2 \ldots \pi_t, \mathsf{close}(f, a)$ such that $\mathsf{State}_A \xrightarrow{tr} \mathsf{State}_B$, the following conditions hold:

- There exists j_1, j_2 such that $1 \leq j_1 < j_2 \leq t, \pi_{j_1} = \mathsf{out}(L_{j_1}, u_{j_1}), \pi_{j_2} = \mathsf{in}(L_{j_2}, Y)$ and $\mathsf{State}_B \models L_{j_1} = f$ and $\mathsf{State}_B \models L_{j_2} = f$.
- Furthermore, there does not exist j such that $j_1 < j < j_2, \pi_j = \mathsf{out}(L_j, u_j)$ and $\mathsf{State}_B \models L_j = a$.

It may be surprising that the above definition does not refer to timing, since distance bounding is enforced by a fast phase where the timing of messages is precisely measured. The above property in fact has been proven to be equivalent to a definition involving timing which we describe informally as follows (this the idea behind a more formal definition [104]): for every trace that contains a distance-bounding claim with fast phase challenge action a and response action b, there are timed occurrences of these events that respect the speed of light. An advantage of having definitions such as Definition 6.3 that do not involve time is that we use tools such as ProVerif and Tamarin to model these properties, since the definition uses the same basic ideas as we use for authentication properties such as recent aliveness.

To understand better why the order of event is sufficient we provide the key intuition that enables us to reduce attacks based on timing attacks to a property concerning the order of events in traces. In Fig. 6.11 we illustrate three scenarios for

6.3 Resisting relay attacks using distance-bounding protocols

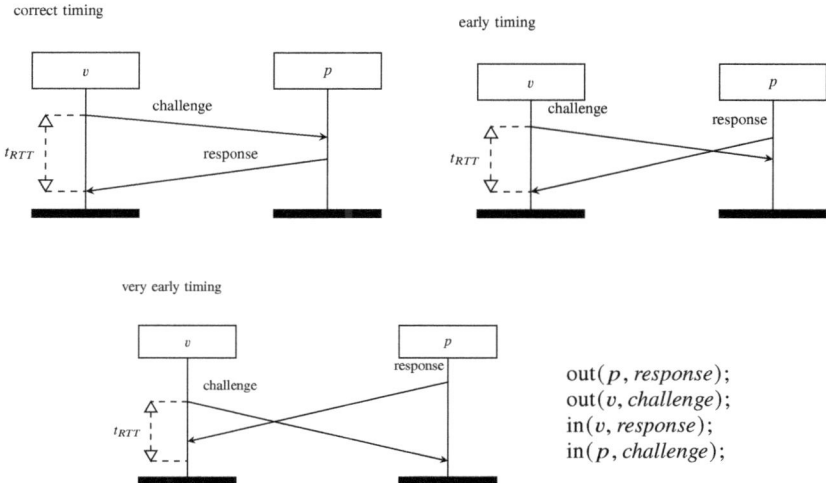

Fig. 6.11 Timing scenarios: *correct timing* is possible when the Prover is near; *early timing* is when the Prover sends a response before the challenge is received in order to pretend to be near; *very early timing* is when the response is sent before the challenge is sent. The trace for very early timing shows no action from the Prover occurs between the actions of the Verifier.

executing a fast phase where the Verifier may believe (in two cases incorrectly) that the Prover is close. The key insight that allows us to use the trace-based definition is as follows. Whenever there is an attack that exploits early timing, there is a attack that exploits very early timing, where the response is sent even before the challenge is sent. Hence, when there is an attack where an attacker fools the Verifier into believing that the Prover is close, there is a trace where the is no send message from the Prover between the Verifier sending a challenge and the Verifier receiving the response.

Coming back to our running example, the Mafia fraud attack on the TREAD protocol is shown in Fig. 6.12.

Using the symbolic notation introduced in this book, the attack trace is as follows.

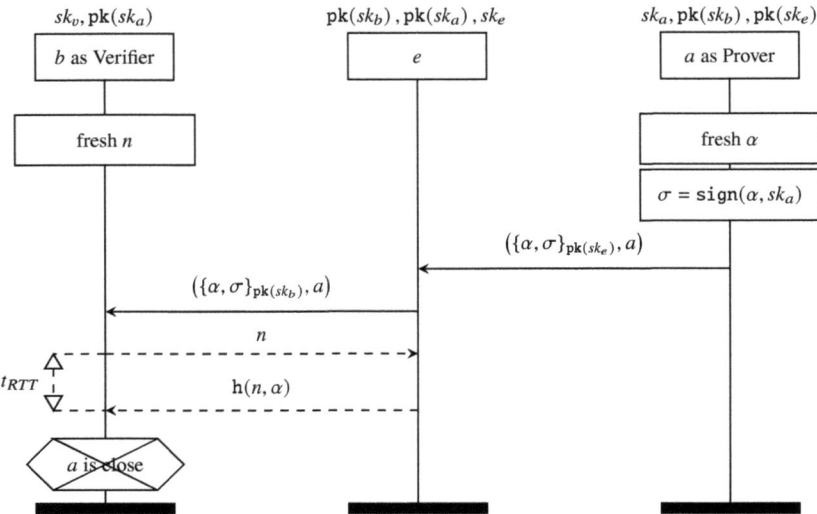

Fig. 6.12 Mafia fraud attack on TREAD Public Key Protocol.

$$\text{State}_{TREAD} \models \begin{array}{l} \langle \text{out}(keys, pk_a) \rangle \\ \langle \text{out}(keys, pk_b) \rangle \\ \langle \text{in}(c, e) \rangle \\ \langle \text{in}(c, \text{pk}(sk_e)) \rangle \\ \langle \text{out}(c, f) \rangle \\ \langle \text{out}(a, u_2) \rangle \\ \langle \text{in}\!\left(c, \left(\{\text{dec}(\text{fst}(u_2), sk_e)\}_{pk_b}, a\right)\right) \rangle \\ \langle \text{out}(f, u_3) \rangle \\ \langle \text{in}(f, \text{h}(u_3, \text{fst}(\text{dec}(\text{fst}(u_2), sk_e)))) \rangle \\ \langle \text{close}(f, a) \rangle \\ \text{true} \end{array}$$

To help read the above, in the final state the active substitution will include the following mappings.

$$\left\{ \begin{array}{l} pk_a \mapsto \text{pk}(sk_a), \\ pk_b \mapsto \text{pk}(sk_b), \\ u_2 \mapsto \left(\{\alpha, \text{sign}(\alpha, sk_a)\}_{\text{pk}(sk_e)}, a\right), \\ u_3 \mapsto n \end{array} \right\}$$

According to our formal definition of Mafia fraud resistance, this trace is an attack on the TREAD protocol. There is a claim close on agent a, stating the fast phase

6.3 Resisting relay attacks using distance-bounding protocols

used the channel f, and between the events that used these channels there is no event by a, as should be the case. This implies a did not respond to the challenge, and as such its distance is not bounded by the round trip time in the fast phase.

6.3.2 Distance hijacking: when the attacker is far away

We consider property where all attackers are assumed to be far away, limiting their ability to interact with the fast phase of the protocol. Such attackers may still attempt to hijack sessions of the protocol by exploiting nearby honest agents.

> Distance hijacking occurs when a Verifier believes that a dishonest Prover e is nearby, even though e is far. Instead, an *honest* Prover p is actually close.

In order to analyse this we assume that all attackers are far away, as suggested in Fig. 6.13, while honest agents may be anywhere. The fact that the attacker is distant limits its ability to actively manipulate messages during the fast phase of a protocol, as we define precisely in this section.

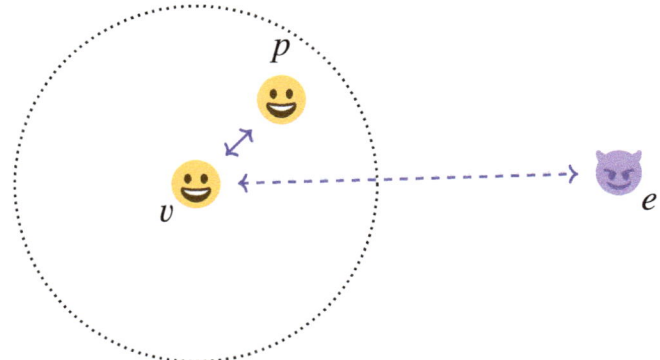

Fig. 6.13 The attacker is far away and only honest verifiers and provers are close.

The distance-hijacking scenario allows an attacker to hijack an honest Prover who is close to pretend that they are close. This scenario is perhaps not so worrying, since, unlike Mafia fraud, it will not enable an attacker to use someone else's credit card or open someone else's car. However, it will allow an attacker to lie about where someone is. For example, by exploiting such a vulnerability, the attacker may be at home with their office access card and open their office door remotely, while never being physically present. This may be exploited to frame someone, for example.

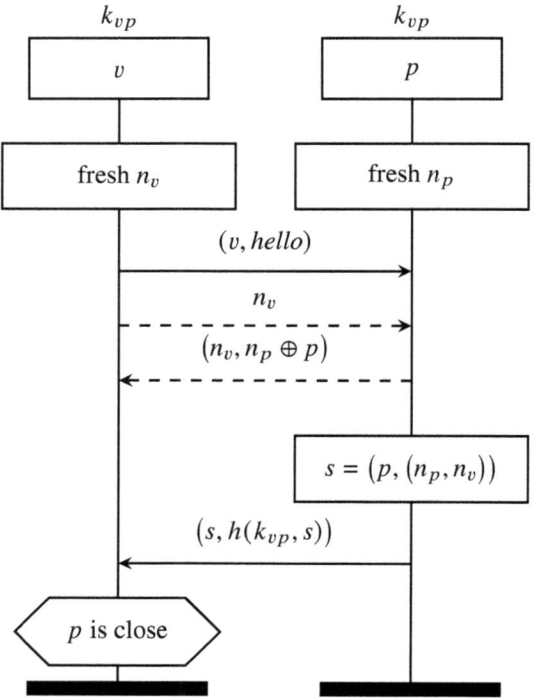

Fig. 6.14 Meadows 2 Protocol.

We illustrate this type of attack on a variant of the Meadows 1 protocol [86], which we will call Meadows 2, to distinguish it from the protocol presented earlier in this chapter.

A distance-hijacking attack on the Meadows 2 protocol is shown in Fig. 6.15. This attack depends on the properties of the XOR operator. To understand how it works, notice that at the beginning b thinks it is starting a session with e, although actually it is a who is acting as the *Prover*. In the response to the fast phase, a sends the message $(n_b, n_a \oplus a)$. This message only contains the identity of a implicitly, and as such b doesn't realize that this message was sent by a. At the end, when e sends the message $(s', h(k_{b,e}, s'))$, b doesn't notice anything wrong, as it is expected that $\text{snd}(\text{fst}(s'))$ is equal to the nonce received during the fast phase, which is actually true because: $\text{snd}(\text{fst}(s')) \oplus e = (n_a \oplus e \oplus a) \oplus e =_E n_a \oplus a \oplus e \oplus e =_E n_a \oplus a$, and this last term was exactly the one received during the fast phase. This trace represents a distance-hijacking attack on the protocol because only a needed to be close to b, while e could be far away and still b thinks the distance between b and e is bounded by the round trip time.

6.3 Resisting relay attacks using distance-bounding protocols

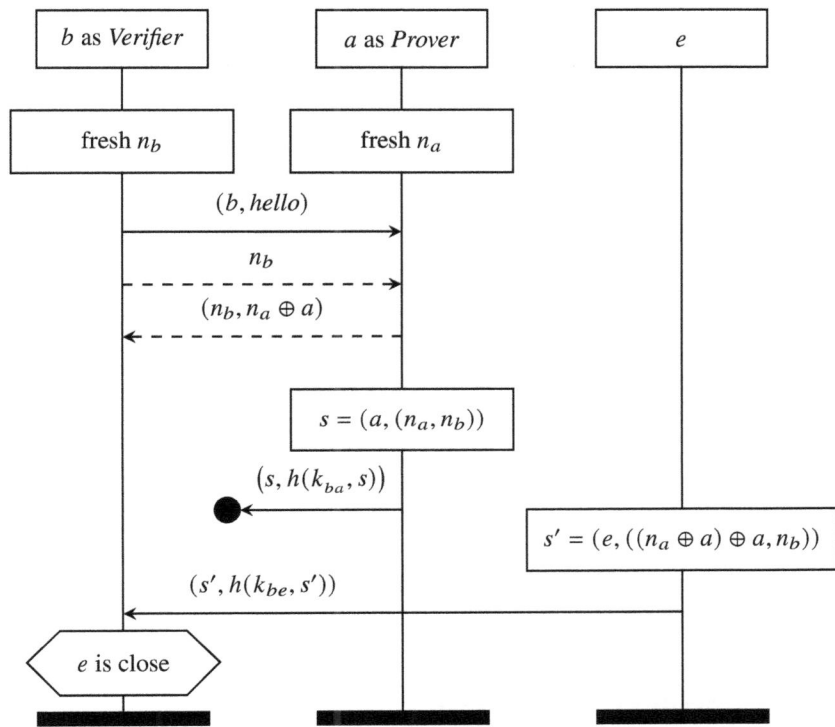

Fig. 6.15 A distance-hijacking attack. Notice having the Verifier b believe the distant attacker e is close (due to the send event from b) is something the protocol designed aims to avoid.

Having formed an intuition for this property, we present a formal definition.

Definition 6.4 (Distance-hijacking attack) A state State_A with respect to a set of honest agents HONEST is *vulnerable to distance hijacking* whenever there exists a trace $tr = \pi_1, \pi_2 \ldots \pi_t, \mathrm{close}(E)$ such that $\mathsf{State}_A \xrightarrow{tr} \mathsf{State}_B$, where $E \notin$ HONEST, and for all j_0, j_1, j_2 such that $1 \leq j_0, j_1, j_2 \leq t$, $\pi_{j_0} = \mathrm{out}(L_{j_0}, f)$, $\pi_{j_1} = \mathrm{out}(L_{j_1}, u_{j_1})$, and $\pi_{j_2} = \mathrm{in}(L_{j_2}, M_{j_2})$, where we have $\mathsf{State}_B \models L_{j_1} = f$ and $\mathsf{State}_B \models L_{j_2} = f$, then, for all $j_1 < i \leq j_2$, if $\pi_i = \mathrm{in}(L_i, M_i)$ then one of the following holds:

- There exists $j < i$ such that $\pi_j = \mathrm{out}(L_j, u_j)$ and $\mathsf{State}_B \models M_i = u_j$.
- For all $j_1 \leq j$, if $\pi_j = \mathrm{out}(L_j, u_j)$, then $u_j \# M_i$.

The above definition contains some nontrivial expressions. The most novel component compared to previous definitions is the restriction on all fast phases. The assumption here is that all fast phases are clearly demarcated by an output action and an input action on a common channel, unique for each session. Between the beginning and end of each fast phase, some input event may occur, including the last input of the fast phase, and possibly other inputs, e.g., inputs by honest nearby provers that may receive quickly an input during the fast phase. For those input messages a distant attacker cannot influence them with full Dolev-Yao capabilities, which explains the two possibilities, below, where at least one must hold.

- The first possibility is that the message was output by an honest agent, who is nearby, unmodified by a distant attacker. Thus, given a message M_i input at i, there must exists a message u_j output at j such that $j < i$ and $\mathsf{State}_B \models M_i = u_j$.
- The second possibility is that the attacker used its knowledge before the start of the fast phase to manufacture a message and sent it early from a distance further away, in time that it arrives during the fast phase. Thus it is impossible that messages output during the fast phase are used by the attacker. Thus, for a message M_i input during the fast phase, if j_1 is the beginning of the fast phase and for some later output u_j at j such that $j_1 \leq j$, we know u_j cannot appear in the recipe for the input M_i, that is $u_j \# M_i$.

The above explains the two clauses that constrain the capabilities of distant attackers, while allowing honest participants to communicate messages during the fast phase.

We need to set up the process threads and network configuration in an appropriate manner, with an honest prover, and an honest verifier that is ready to connect to honest provers (coinciding with the set of honest provers above), and dishonest provers. In the following, the set of honest agents is $\{a, b\}$ only.

$$\mathsf{State}_{MEADOWS2} \triangleq \mathsf{new}\ k_{ba};\ (\\
(!\ Prover(c, a, k_{ba}, b))\ |\\
(!\ in(c, e);\\
\quad \text{if}\ e = a\ \text{then}\\
\quad\quad Verifier(c, b, k_{ba}, a)\\
\quad \text{else}\\
\quad\quad in(c, k_{be});\ Verifier(c, b, k_{be}, e)\ ))$$

(INITIAL NETWORK CONFIGURATION)

For the Meadows 2 protocol the roles are as follows.

$$Prover(c, p, k_{vp}, v) \triangleq \mathsf{new}\ n_p;\\
\quad in(c, x);\\
\quad \text{if}\ x = (v, hello)\ \text{then}\\
\quad in(c, n_v);\\
\quad out(c, (n_v, n_p \oplus p));\\
\quad \text{let}\ s = (p, (n_p, n_v))\ \text{in}\\
\quad out(c, (s, h(k_{vp}, s)));;\ 0$$

In the Verifier thread below, the fast phase is marked by a fresh channel within each session that is picked out in the "close" claim actions. The identity of the agent

6.3 Resisting relay attacks using distance-bounding protocols

believed to be close is identified in the claim only. This allows us to check that the claim is never reached if the agent identified is dishonest. Since we range over all fast phases, we need not identify a specific fast phase in the claim.

$Verifier(c, v, k_{vp}, p) \triangleq$ new f; out(c, f);   Channel for fast phase.
                  new n_v;
                  out$(c, (v, hello))$;
                  out(f, n_v);   Beginning of fast phase.
                  in(f, x);   End of fast phase.
                  if $\mathtt{fst}(x) = n_v$ then
                  in(c, y);
                  if $\mathtt{fst}(y) = (p, (\mathtt{snd}(x) \oplus p, n_v))$ then
                  if $\mathtt{snd}(y) = \mathsf{h}(k_{vp}, \mathtt{fst}(y))$ then
                  close(p)

Using the notation introduced in this book, an attack trace is as follows.

$\text{State}_{MEADOWS2} \models \langle \text{in}(c, e) \rangle$ where
                   $\langle \text{in}(c, k_{be}) \rangle$
                   $\langle \text{out}(c, f) \rangle$
                   $\langle \text{out}(c, u_2) \rangle$ new n_a, n_b;
                   $\langle \text{in}(c, (b, hello)) \rangle$ $u_2 \mapsto (b, hello)$
                   $\langle \text{out}(f, u_3) \rangle$ $u_3 \mapsto n_b$
                   $\langle \text{in}(c, u_3) \rangle$ $u_4 \mapsto (n_b, n_a \oplus a)$
                   $\langle \text{out}(c, u_4) \rangle$ $s' = (e, (\mathtt{snd}(u_4) \oplus e, u_3))$
                   $\langle \text{in}(f, u_4) \rangle$
                   $\langle \text{in}(c, (s', \mathsf{h}(k_{be}, s'))) \rangle$
                   $\langle \text{close}(e) \rangle$
                   true

In the above, you can see that there is one fast phase on fresh channel f. That fast phase consists of messages named u_3 and u_4 that are sent and received by the honest prover and verifier, unmodified. Thus the first clause at the end of Def. 6.4 applies, making these legitimate messages. Put simply, a regular honest fast phase between honest agents happens. Yet, at the end, the verifier believes that the attacker is close, which is not the case.

6.3.3 Practice in verifying distance-bounding protocols

You have now the tools to address the question on Mafia fraud formally.

Exercise 6.3 This question concerns the distance-bounding protocols in Figures 6.16 and 6.17.

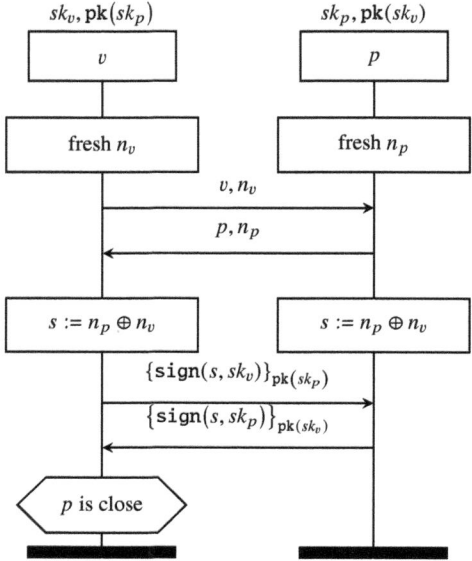

Fig. 6.16 An example distance-bounding protocol for Exercise 6.3.

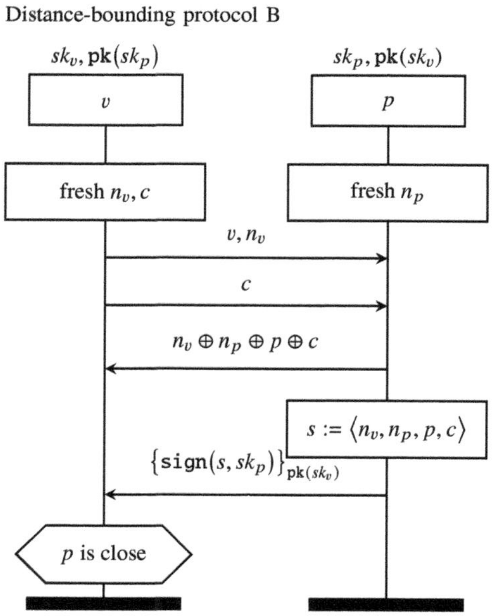

Fig. 6.17 Another example distance-bounding protocol Exercise 6.3.

6.3 Resisting relay attacks using distance-bounding protocols

1. Indicate which message(s) in the MSC for the distance-bounding protocol A in Fig. 6.16 correspond to the fast phase of this protocol.
2. The distance-bounding protocol A is vulnerable to Mafia-fraud, hence the claim "p is close" can be violated even when p is honest and not close. Give a trace that demonstrates this.
3. Changing one message is sufficient to guarantee that the distance-bounding protocol A is Mafia-fraud resistant. Give an example of a fixed version of the protocol, and an informal explanation of why your fix works.
4. In distance hijacking, we are interested in attacks on the claim "p is close" even when p is not an honest agent. Give an attack on the distance-bounding protocol B in Fig. 6.17 that violates this property.

For further practice, analyse the claims in the distance-bounding protocol in the next exercise.

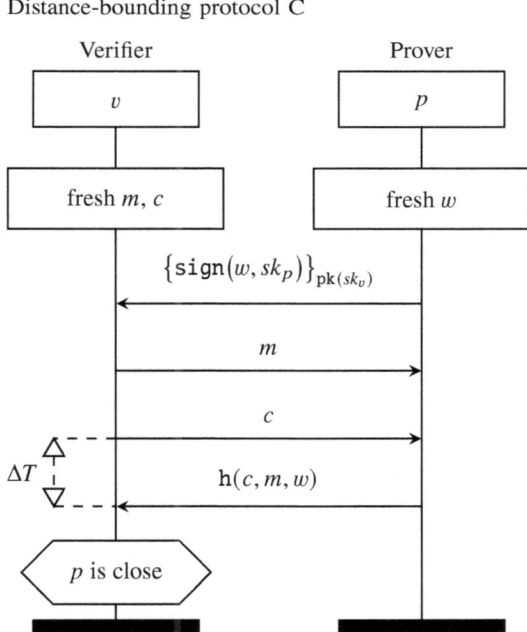

Fig. 6.18 An example distance-bounding protocol for Exercise 6.4.

Exercise 6.4 Consider the distance-bounding protocol shown in Fig. 6.18. The fast phase is denoted by a single challenge/response round (the two marked with the time-measure sign ΔT) instead of several rounds.

1. Find an attack in which an adversary (close to the verifier) tricks the verifier into believing that an honest prover is close. Model the protocol, threat model and attack formally.
2. Is there a distance-fraud attack against this protocol?

6.4 Tooling for aliveness and distance bounding

We cover tooling for recent aliveness and Mafia fraud here together. Both properties make use of features allowing the temporal order of events to be constrained. Temporal queries are relatively new to ProVerif [29], making this section perhaps the most experimental of this book. This explains why tooling for recent aliveness, is delayed until this later stage of this book. Due to the experimental nature of temporal queries, tooling may evolve in the near future, however, the temporal queries presented here are logically accurate and we expect to survive evolution. Note that the tool Tamarin, covered in a related book in this series [22], has long supported temporal queries.

6.4.1 Tooling for aliveness

We have previously mentioned the automatic protocol verification tool Scyther. This tool can also model weak aliveness claims. Below you can find the specification of $Hello_1$ in Scyther.

const Hello;

protocol *hello1*(I, R) {
    role I {
        send_1$(I, R, \{$Hello$\}sk(I))$;
        recv_2$(R, I, \{$Hello$\}sk(R))$;
        claim_3$(I, $ Alive$, R)$;
    }

    role R {
        recv_1$(I, R, \{$Hello$\}sk(I))$;
        send_2$(R, I, \{$Hello$\}sk(R))$;
    }
}

    The Alive claim above is implemented in Scyther as weak aliveness only. Hence Scyther establishes that the above claim holds, without discovering an attack that would be expected for recent aliveness. You may also derive a Scyther specifications

6.4 Tooling for aliveness and distance bounding

for $Hello_0$, and analyse the manner in which the tool describes the trace of the attack found on $Hello_0$.

In ProVerif, recent aliveness can be specified by making use of queries that refer to when in a trace a message occurs. We make use of the following events and queries to annotate a protocol specification with the appropriate information.

event started(bitstring).
event sentMessage(bitstring).
event alive(bitstring, bitstring).

The query for recent aliveness above makes use of a temporal query. The annotations on events and constraints ensure that some event at timestamp $t1$ certainly happened before an event at timestamp $t2$. Specifically, the session of the agent making the authentication claim started before a message was sent by the communicating partner, and hence the communicating partner was alive at least that recently.

query f: bitstring, r: bitstring, $t1$: $time$, $t2$: $time$;
  event(alive(f, r)) $\wedge$ **event**(started(f))@$t1$
$\implies$
  (**event**(sentMessage(r))@$t2 \wedge t1 < t2$).

The following events appear in the query above and roles below:

- Event started(f) is positioned to have necessarily occurred if the initiator has started communicating in a session. Each session is identified uniquely by a fresh name f, not used in the protocol.
- Event sentMessage(r) that has necessarily occurred if the responder has received and sent some message.
- Event alive(f, r) that occurs if the initiator completes a session, uniquely identified by f, that is intended to involve an honest responder r.

Let us consider the simple protocol with roles specified as follows, based on MACs.

const hello: bitstring.

let *Initiator*(c: channel, r: bitstring, mk: bitstring) =
  new f: bitstring;
  event started(f);
  out(c, hello);
  in(c, y: bitstring);
  if h((hello, mk)) = y **then**
  event alive(f, r).

let *Responder*(c :channel, r, mk:bitstring) =
  in(c, $hello'$: bitstring);
  if $hello'$ = hello **then**
  event sentMessage(r);

out(c, h((hello, *mk*))).

The threat model for symmetric key cryptography is set up as normal, as you can see below.

free c: channel.
free b: bitstring.
process new *mk*: bitstring; (
   (!*Initiator*(c, b, *mk*)) |
   (!*Responder*(c, b, *mk*)))

The above code should, in principle, allow ProVerif to discover an attack on recent aliveness from the perspective of the initiator. Surprisingly, this may be the simplest protocol and property seen and, at the time of writing, we need to add the prefix **inj–** to events for an attack to be discovered. An attack on the property with injectivity added could mean that two verifier sessions are reusing a message in the same prover session. By inspecting the trace you can see if swapping events is possible to move one verifier session after another, in which case there is also an attack on the proper definition of recent aliveness in Def. 6.2 that does not enforce injectivity. This is a manual observation.[4] As stated at the beginning of this section, support in ProVerif for temporal queries is relatively recent, so this observation may evolve.

The query for weak aliveness in ProVerif, below, is simpler, requiring only that some event occurred without constraining when it occurred.

query f: bitstring, b: bitstring; *(* weak aliveness *)*
  event(alive(f, b)) $\implies$ **event**(sentMessage(b)).

Using the above query, the above code also proves that weak aliveness holds. However, if we allow both agents to play either role, as in the following, also weak aliveness will fail.

free c: channel.
free a, b: bitstring.
process new *mk*: bitstring; (
   (!*Initiator*(c, b, *mk*)) |
   (!*Responder*(c, b, *mk*)) |
   (!*Initiator*(c, a, *mk*)) |
   (!*Responder*(c, a, *mk*)))

The above models a variant of weak aliveness where agents may assume multiple roles. The event in the right hand side of the weak aliveness query insists on the agent who acts. Since one agent may play either role, they may be talking to themselves, and the other agent was never alive. ProVerif will indeed find an attack where only one

---

[4] We are grateful to Bruno Blanchet who confirmed that this represents the state of the art in 2025.

agent is used to complete the protocol. Hence the relevant sentMessage event never appears and so the query fails, as expected. This shows how sensitive authentication can be to small variations in the threat model. A related book [46], analyses further variants of aliveness.

6.4.2 Practice in Tooling for Mafia fraud

The exercises in this section give the reader experience with using temporal queries. To express Mafia fraud resistance (Def. 6.3) in ProVerif, we use a query constraining the order of events. The query should express that an event from an honest communicating partner occurs between the beginning and end of the fast phase.

Earlier, we claimed that the Meadows 1 protocol in Fig. 6.8 is Mafia-fraud resistant. ProVerif can verify this claim as we break down here. Meadows 1 makes use of a hash function only, with no equations.

fun h(bitstring): bitstring.

To express Mafia fraud, we make use of events that mark the beginning and end of the fast phase – startff and endff which share a fresh name, serving the role of the fast channel in Sec. 6.3. There is also event close to mark the end of the protocol identifying the fast phase and the agent that should be near. Event outprover marks outputs of the prover, parameterised on the prover's identity.

event close(bitstring, bitstring).
event startff(bitstring).
event endff(bitstring).
event outprover(bitstring).

The above events are then assembled as follows, to assert that some event of the prover happened during the relevant fast phase if the verifier reaches the end of the protocol successfully.

query f: bitstring, p: bitstring, i: *time*, j: *time*, k: *time*;
  event(close(p,f))$\wedge$
  event(endff(f))@$k\wedge$
  event(startff(f))@i
$\implies$
  (**event**(outprover(p))@$j \wedge i < j \wedge j < k$).

As normal, we define the roles of the protocol, but annotated with the above events. In the following, a new name f is created to tie all the events of the verifier together in this run of the protocol.

let *verifier*(c: channel, v: bitstring, k: bitstring, p: bitstring) =
  out(c, v);
  new f: bitstring;
  event startff(f);
  new n: bitstring;
  out(c, n);
  in(c, (n': bitstring, m: bitstring, p': bitstring));
  event endff(f);
  if $n = n' \wedge p = p'$ **then**
  in(c, y: bitstring);
  if $y = \mathsf{h}((k, p, m, n))$ **then**
  event close(p, f) .

An event is positioned immediately before the outputs of messages, which is used to detect whether the prover is active during the fast phase.

let *prover*(c: channel, p: bitstring, k: bitstring, v: bitstring) =
  in(c, v': bitstring);
  if $v = v'$ **then**
  in(c, n: bitstring);
  new m: bitstring;
  event outprover(p);
  out(c, (n, m, p));
  out(c, $\mathsf{h}((k, p, m, n))$) .

As normal, for symmetric key protocols we don't need to consider open network connections and so the network can be defined as follows.

free a, b: bitstring.
free c: channel.
process new k: bitstring; (
    (!*verifier*(c, b, k, a)) |
    (!*prover*(c, a, k, b)))

Exercise 6.5 The following showcases some quite experimental capabilities of ProVerif for analysing Mafia fraud.

1. Use the ProVerif code above to check Mafia-fraud resistance for Meadows 1 protocol from Fig. 6.8. If the protocol cannot be proven, annotate events in the above query with the **inj−** prefix. Justify that a proof on the injective variant of Mafia fraud also a valid proof of the original non-injective definition.
2. Consider the public key version of the TREAD protocol in Fig. 6.10. Provide a ProVerif implementation of the protocol. Recall that this example is presented in detail throughout Sec. 6.3.1. You should reuse the open network model presented

6.4 Tooling for aliveness and distance bounding

in Sec. 6.3.1, and adapt the threads for roles so that events are used as illustrated here in Sec. 6.4.2.

Find whether ProVerif discovers an attack Mafia fraud on this protocol with or without making the query injective (similar to the previous question). If an attack is discovered using injectivity, can the attack itself be transformed into an attack on Mafia fraud manually by swapping events.

The above exercise covers Mafia fraud but does not cover distance hijacking. While modelling distance hijacking is possible in ProVerif and Tamarin, using tools to find proofs and attacks is beyond the scope of this book. We also did not cover tooling for protocols featuring XOR. At the time of writing, XOR is best handled using the tool Maude-NPA [54]. We anticipate more convergence between tools as advances in one tool inform advances in another tool.

Chapter 7
Privacy

Abstract Unlinkability, as standardised by ISO 15408, is formalised as an equivalence problem between a normal network and one where it is impossible to reuse identities between sessions. A notion of process equivalence, bisimilarity, is introduced along with its characteristic modal logic. The modal logic is applied to analyse known vulnerabilities in various implementations of ePassport protocols. Unlinkability of EMV is also analysed and exercises consider unlinkability of further protocols. Finally, a protocol is developed that satisfies unlinkability and also various authentication properties, plus a strong confidentiality property – forward secrecy. The verification of all properties of this showcase protocol is automated.

7.1 Introduction

Security properties such as secrecy and agreement involve data transmitted during the execution of a protocol being revealed or manipulated. There are also privacy properties that define more subtle ways for information to leak from its intended context into another unintended context. Even if secrecy and agreement hold, there may be privacy leaks.

The Common Criteria for Information Technology Security Evaluation ISO/IEC 15408 defines privacy properties desirable for information systems [6]. They key privacy properties covered by that standard are the following.

I.1 Anonymity – the client is not allowed to know the identity of the user.

I.2 Pseudonymity – a user may use a resource or service without disclosing their identity, but can still be accountable for that use.

I.3 Unlinkability – a user may make multiple uses of resources or services without others being able to link these uses together.

I.4 Unobservability – a user may use a resource or service without others, especially third parties, being able to observe that the resource or service is being used.

Firstly consider the arguably the strongest property: unobservability. As an example of this property, the OCR session between a QR code and a phone is unobservable to an attacker snooping on wireless communications. In contrast, even if a protocol for a contactless card is secure in every sense we have covered and satisfies anonymity and unlinkability, the fact that a transaction occurred is observable to the same attacker snooping on wireless communications. Even if no information about specific transactions is revealed, some information, e.g., global statistics about the numbers of transactions performed and their location may be revealed. Attacks exploiting observability, that just observe transactions and gather overall statistic are more likely to be in scope of network traffic methods rather than the symbolic methods in this book.

After unobservability, the next strongest property is unlinkability. Suppose that for some protocol there is no way of telling which agent was executing a protocol. We may, none-the-less, be able to tell that the same agent executed the protocol more than once. That is, it may be possible to link two sessions of the protocol involving the same agent and use that information to track behaviours of agents.

In contrast to unlinkability, an attack on anonymity would allow the identity of the agent to be revealed, thereby making a specific agent traceable. For example, many of the classic protocols we have studied earlier involve explicit identities for agents or a public key associated with an agent, and that information can often be used to learn relatively easily who was executing a session. Indeed many protocols were not designed with privacy in mind, for example the EMV standard transmits the PAN (the account number of the cardholder) in cleartext, which is strongly identifying information – thus permitting attacks on anonymity and hence unlinkability.

7.2 The strong privacy property: unlinkability

We will focus on unlinkability in this section, which is perhaps the strongest privacy we can expect when message are sent and received over a network on which an attacker may intercept communications. If unlinkability holds, then anonymity usually also holds, since if the identity of an agent involved can be revealed for every session, then two session may be linked simply by seeking two sessions involving the same identity. This makes unlinkability a good benchmark for establishing the privacy of a protocol.

We can express unlinkability as an equivalence problem between two networks, as suggested below in the case of ePassports.

7.2 The strong privacy property: unlinkability

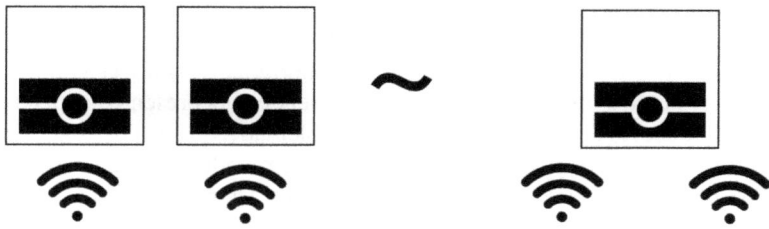

In the above, we compare an *idealised world* on the left, where every reader session is with a fresh ePassport, to a system on the right, where, as in the *real world*, the same ePassport can be used across multiple sessions. If these two networks are equivalent from the perspective of an attacker observing communications, then the attacker cannot do anything to tell whether the same ePassport was used across multiple distinct sessions. That is, the sessions are unlinkable.

Privacy properties can typically be formulated as *indistinguishability properties* since they are often defined as the indistinguishability between two worlds: the real one, where the system operates, and an idealised one, where the desired property holds trivially. This contrasts significantly with the *reachability properties* we have studied so far, which are defined in terms properties that hold upon reaching a state.

To model this ePassport unlinkability problem, we need the following:

- We must clarify how to model these two scenarios we are comparing as networks.
- We also must clarify what the symbol $\sim$ means.

We explain first how to model the networks, which reuses existing methods we are familiar with by now. The definition of $\sim$ will require additional machinery.

As normal, we model the roles of a protocol, describing the message inputs and outputs, freshly generated names and conditions. For example, below is a model of a French version of the BAC protocol used in ePassports.

$$P_{Fr}(c, ke, km) \triangleq \mathsf{new}\ nt;\ \mathsf{out}(c, nt);\mathsf{in}(c, y);$$
$$\quad \mathsf{if}\ \mathsf{snd}(y) = \mathsf{mac}(\mathsf{fst}(y), km)\ \mathsf{then}$$
$$\quad \quad \mathsf{if}\ nt = \mathsf{fst}(\mathsf{snd}(\mathsf{dec}(\mathsf{fst}(y), ke)))\ \mathsf{then}$$
$$\quad \quad \quad \mathsf{new}\ kt;$$
$$\quad \quad \quad \mathsf{let}\ m = \{(nt, (\mathsf{fst}(\mathsf{dec}(\mathsf{fst}(y), ke)), kt))\}_{ke}\ \mathsf{in}$$
$$\quad \quad \quad \mathsf{out}(c, (m, \mathsf{mac}(m, km)));;\ 0$$
$$\quad \quad \mathsf{else}\ \mathsf{out}(c, e_{nonce});;\ 0$$
$$\quad \mathsf{else}\ \mathsf{out}(c, e_{mac});;\ 0$$

In addition to what was explained in Sec. 2.2, the protocol features two different error messages. The error message e_{mac} occurs when the condition $\mathsf{snd}(y) = \mathsf{mac}(\mathsf{fst}(y), km)$ fails. Thus if the message-MAC pair sent match according to the expected key *km*, then that error message will not occur. The other error message, e_{nonce}, occurs if the above MAC test passes but the message received, when decrypted is found not to contain the nonce sent out originally by the ePassport. Thus, in a failed authentication session, there are different errors presented depending on the reason for aborting, which we will see is problematic.

An informal MSC is presented in Fig. 7.1 to help understand the above formal definition of the French ePassport. We note that this accurately models the French ePassport in 2010 [13]. Modern passports should use the PACE protocol that mitigates this problem by stipulating a specific error message for all failure modes.

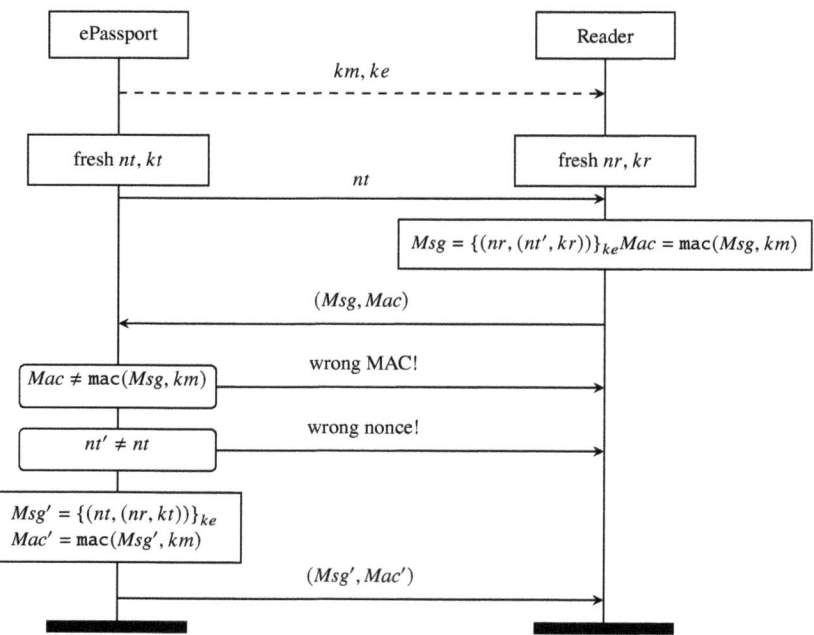

Fig. 7.1 Message sequence chart for BAC as in the French ePassport.

The first two messages of an ePassport reader that verifies ePassports are defined according to the following thread.

$$V(c, ke, km) \triangleq \text{in}(c, nt); \text{new } nr, kr;$$
$$\text{let } m = \{(nr, (nt, kr))\}_{ke} \text{ in}$$
$$\text{out}(c, (m, \text{mac}((m, km))));$$

7.2 The strong privacy property: unlinkability

We can now describe a system, modelling realistic usage of ePassports, where there may be many ePassports and each of them may be used in many sessions.

$$Sys_{Fr} \triangleq \text{!new } ke, km; !(\text{ new } c; \text{out}(v,c); V(c,ke,km) \mid \\ \text{new } c; \text{out}(p,c); P_{Fr}(c,ke,km))$$

Observe that each time a fresh ke and km are generated, which models the creation of an ePassport manufactured so that it responds to those keys only. The use of replication before ke and km is what models many ePassports being manufactured, while the replication after ke and km shows that ePassport can be used many times.

As when defining agreement, we generate a fresh channel for each ePassport and reader session and output it so that anyone may use the channel. This means we have a unique channel for each endpoint. This is a realistic assumption, since it models setting up a channel using the underlying wireless network protocol before starting the authentication protocol.

The idealised world where unlinkability holds by definition can be described by removing only the innermost replication, thereby preventing an ePassport from being involved in more than one session. The precise specification is described below.

$$Spec_{Fr} \triangleq \text{!new } ke, km; (\text{ new } c; \text{out}(v,c); V(c,ke,km) \mid \\ \text{new } c; \text{out}(p,c); P_{Fr}(c,ke,km))$$

The question is now whether Sys_{Fr} and $Spec_{Fr}$ are equivalent from the perspective of an attacker intercepting outputs and producing inputs using the outputs intercepted.

What it means for two networks to be equivalent is a deep question, since there are many slightly different notions of equivalence for processes. We present a well-established answer to this question next.

7.2.1 Bisimilarity: checking there is no distinguishing strategy

Bisimilarity is an equivalence relation that equates processes that are indistinguishable from the perspective of the attacker, even if they may be internally different in some way that the attacker cannot detect. An equivalence relation, is simply a binary relation with the properties you would expect for an equality: it is transitive (if $A \sim B$ and $B \sim C$ then $A \sim C$), symmetric (if $A \sim B$ then $B \sim A$) and reflexive ($A \sim A$).

The definition of bisimilarity, the relation $\sim$ below, is concise, but will require some explanation.

Definition 7.1 A *bisimulation* is a symmetric relation S such that whenever $A \, S \, B$:

- If $A \xrightarrow{\pi} A'$, there exists B' such that $B \xrightarrow{\pi} B'$ and $A' \, S \, B'$.
- For all messages M and N, if $A \models M = N$ then $B \models M = N$.

We say that A and B are bisimilar, written $A \sim B$, whenever there exists a bisimulation S such that $A \, S \, B$.

To establish that two states are bisimilar we construct a relation over states called a *bisimulation*. The first clause above ensures that the bisimulation relation contains all states that either process passes through and equivalence of states is maintained throughout execution. In a pair of states during execution, whenever a state can perform an input or output action, π in the above definition, then the other process can perform the same action π, and we stay within the relation. The second clause ensures that the knowledge accumulated in the active substitution at any step is equivalent from the perspective of the attacker for both states being compared.

We can pose the problem of unlinkability of the French ePassport, by attempting to prove $Sys_{Fr} \sim Spec_{Fr}$, which would involve constructing a large bisimulation relation. For the French implementation of the BAC protocol, it turns out that the converse holds.

Theorem 7.1 $Sys_{Fr} \not\sim Spec_{Fr}$.

As for proving secrecy or authentication properties, it is easier to disprove something than to prove it. Disproving amounts to describing an attack while proving requires more sophisticated reasoning.

Using a marginal extension of the modal logic that we have already been using to describe attacks, we can describe a trace of actions that Sys_{Fr} can perform, but $Spec_{Fr}$ cannot match, as described below. The only additional feature we employ is that we can make use of conjunction to declare that e is e_{nonce} and u_2 is not e_{mac} (and hence u_2 must be a legitimate ciphertext given that this sequence of messages can never lead to u_2 being e_{nonce} since the correct nonce was fed to the reader). In the following, $\not\models$ means that attempting to prove with $\models$ fails.

$$Sys_{Fr} \models \langle\text{out}(p, c_1)\rangle \qquad Spec_{Fr} \not\models \langle\text{out}(p, c_1)\rangle$$
$$\langle\text{out}(v, c_2)\rangle \qquad \langle\text{out}(v, c_2)\rangle$$
$$\langle\text{out}(c_1, nt_1)\rangle \qquad \langle\text{out}(c_1, nt_1)\rangle$$
$$\langle\text{in}(c_2, nt_1)\rangle \qquad \langle\text{in}(c_2, nt_1)\rangle$$
$$\langle\text{out}(c_2, u_1)\rangle \qquad \langle\text{out}(c_2, u_1)\rangle$$
$$\langle\text{in}(c_1, u_1)\rangle \qquad \langle\text{in}(c_1, u_1)\rangle$$
$$\langle\text{out}(c_1, u_2)\rangle \qquad \langle\text{out}(c_1, u_2)\rangle$$
$$\langle\text{out}(p, c_3)\rangle \qquad \langle\text{out}(p, c_3)\rangle$$
$$\langle\text{out}(c_3, nt_2)\rangle \qquad \langle\text{out}(c_3, nt_2)\rangle$$
$$\langle\text{in}(c_3, u_1)\rangle \qquad \langle\text{in}(c_3, u_1)\rangle$$
$$\langle\text{out}(c_3, e)\rangle \qquad \langle\text{out}(c_3, e)\rangle$$
$$(e = e_{nonce} \wedge u_2 \neq e_{mac}) \qquad (e = e_{nonce} \wedge u_2 \neq e_{mac})$$

As indicated above, the given formula is satisfied by the system on the left, but not the specification on the right. This means that the trace represents something that we can do in the real world, that should not be possible if unlinkability holds. The formula describes a strategy for attacking unlinkability of the French ePassport.

The attack strategy described by the trace is depicted by the MSC in Fig. 7.2. In the attack strategy, two ePassport sessions are started involving the same keys, hence are sessions with the same ePassport. The first ePassport starts talking to a

7.2 The strong privacy property: unlinkability

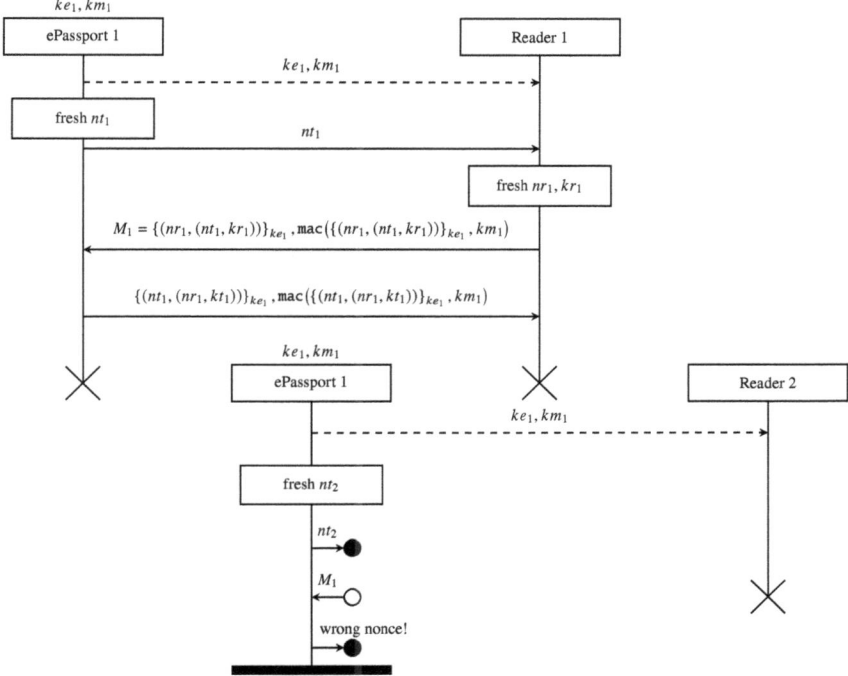

Fig. 7.2 Attack on French ePassports.

reader loaded with the correct keys, and the response of the reader is intercepted by the attacker (e.g., by snooping at a distance of say 20m away on a legitimate session involving an ePassport). That legitimate session runs to completion, passing messages unaltered between the ePassport and reader. The final message from the ePassport is thereby a response, u_2, which is not an error message. If the keys were different $u_2 = e_{mac}$, and hence the test on u_2 ensures that indeed the systems is such that the ePassport and reader share the same keys up to that point.

We are now interested in replaying the first response of the reader, u_1 in the formula above, in what follows. In the second ePassport session, the attacker ignore the new challenge nt_2, and then replays u_1, from the previous reader session, as the response to the challenge. In Fig. 7.2, observe that u_1 corresponds to M_1, which is sent in the first session and input to the second session. If the MAC key km_1 is the same as previously, the MAC test will pass. However, since the nonce is not the nonce generated by the second session, we will receive some message e which we can test is equal to e_{nonce}. This is a legitimate run of the system, hence the formula with the test $e = e_{nonce}$ is satisfied by the system.

In contrast to the above, the idealised specification performs these actions the final message will be e_{mac}, rather than e_{nonce}, since the mac keys are expected to be different in any two sessions. Recall that, in the idealised specification, no two sessions can involve an ePassport with the same keys.

In summary, the formula above is a proof that there is no bisimulation relating Sys_{Fr} and $Spec_{Fr}$. All diamond modalities in the formula represent something that Sys_{Fr} can do, and hence, in any bisimulation, $Spec_{Fr}$ should be able to match all those steps. Indeed $Spec_{Fr}$ can match any action of Sys_{Fr}, so this is not where constructing a bisimulation fails. It is the second clause of the definition of a bisimulation, ensuring that all equalities between messages are preserved, that is violated, in the final state reached. Note that the symmetry of a bisimulation ensures that equalities are preserved in both directions, and hence also inequalities must also be preserved. While the system can reach a state where $e = e_{nonce}$ and $e \neq e_{nonce}$, in all states that the idealised specification can reach by the same actions where $e \neq e_{mac}$, it is impossible that $e = e_{nonce}$ holds.

7.2.2 Hennessy-Milner and the modal logic classical $\mathcal{FM}$

We summarise the rules of the modal logic we have accumulated. By defining conjunction and negation in full, we obtain the modal logic classical $\mathcal{FM}$.

$$\text{new } \mathbf{x}; [\sigma \mid P] \models M = N \quad \text{iff } M\sigma =_E N\sigma \text{ and } \mathbf{x} \mathbin{\#} M, N$$
$$A \models \langle \pi \rangle \phi \qquad \text{iff there exists } B \text{ s.t. } A \xrightarrow{\pi} B \text{ and } B \models \phi.$$
$$A \models \phi_1 \wedge \phi_2 \qquad \text{iff } A \models \phi_1 \text{ and } A \models \phi_2.$$
$$A \models \neg \phi \qquad \text{iff } A \models \phi \text{ fails.}$$

The above semantics define a classical logic. Hence, De Morgan properties define other connectives: for instance, disjunction can be defined using conjunction and negation as follows $\neg(\neg\phi \wedge \neg\psi) = \phi \vee \psi$.

As illustrated above for the BAC protocol implemented in the French ePassport, we were able to explain why two processes are not bisimilar using a modal logic formula that holds for one network but not the other. In fact this property holds in general, in that any two bisimilar processes satisfy exactly the same formulas, as expressed below.

Theorem 7.2 (Hennessy-Milner property [68]) *$A \sim B$ whenever, for all ϕ, we have $A \models \phi$ iff $B \models \phi$.*

Thus the converse of the above is that two networks are not bisimilar whenever there exists a formula that holds for one process but not the other. Since we have expressed unlinkability in terms of bisimilarity, our modal logic therefore can express any attack on unlinkability detectable in the model presented. This is a classic theorem, first established by Hennessy and Milner for a much simpler process calculus [65].

7.2 The strong privacy property: unlinkability

7.2.3 Privacy in card payments

Recall from Sec. 2.3, that the EMV protocol violates all privacy properties, since the card number PAN is transmitted in cleartext, as are transaction details. For contactless cards this is particularly problematic since a passive attacker may learn who is involved in a transaction and information such as values of purchases by eavesdropping on wireless communication in wide radius given the right equipment [53].

To address this problem EMVco proposed a 2nd Generation key agreement protocol [5] that aimed to produce a key that would be used to encrypt transaction details and to establish unlinkability. This proposal is presented in Fig. 7.3. The protocol involves a Diffie-Hellman handshake with an extra nonce, which is intended to blind a long-term key used by the card, hence its name Blinded Diffie-Hellman (BDH). This is now incorporated into Kernel 8 of the EMV standard [52].

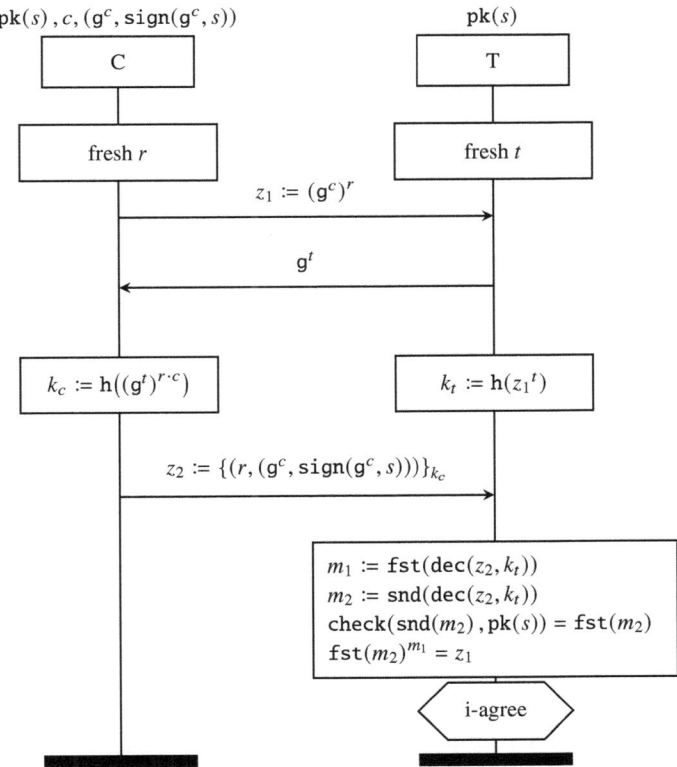

Fig. 7.3 Blinded Diffie-Hellman: handshake for EMV 2nd gen protocol.

Before presenting the protocol, we introduce some notation for expressing Diffie–Hellman handshakes and simplify the notation for the EMV protocol presented earlier. Let s and c be the secret keys of the payment system and the card, respectively. Then, $\text{pk}(s)$ is the public key of the Payment System that the terminal stores and uses to verify any signature made with s. Further, g^c is the long-term public key of the card, i.e., the (public) generator of the Elliptic Curve Group g raised to the power of c. This is standard for Elliptic-curve Diffie-Hellman, where elements like c, called scalars, play the role of a secret key, and elements of the form g^c play the role of public keys.

We now extend the equational theory E to AE, with the following equations for multiplication and exponentiation.

$$M \cdot N =_{\text{AE}} N \cdot M \qquad (M \cdot N) \cdot K =_{\text{AE}} M \cdot (N \cdot K) \qquad K^{M \cdot N} =_{\text{AE}} \left(K^N\right)^M$$

We assume that the EMV chain of certificates comprises just one certificate $(\text{g}^c, \text{sign}(\text{g}^c, s))$, i.e., that the Payment System directly issues and signs cards. The protocol runs as illustrated in Fig. 7.3. The card starts the communication by sending its public key g^c blinded with a fresh scalar r to the terminal. In response, the terminal sends ephemeral public key g^t to the card. This is enough to establish a common secret key $k_c = k_t$. The card uses this key to encrypt the authentication data: blinding scalar r and the certificate $(\text{g}^c, \text{sign}(\text{g}^c, s))$. Finally, the terminal verifies the received certificate by checking the signature against the public key of the payment system $\text{pk}(s)$ and checks that g^c blinded with r coincides with the first message z_1 received from the card. Upon success, the terminal authenticates the card and is ready to continue with the transaction on the encrypted channel.

The card can now be modelled by the following thread.

$$\begin{aligned}
C(s, c, ch) \triangleq \ & \text{new } r; \\
& \text{out}(ch, (\text{g}^c)^r) \\
& \text{in}(ch, y); \\
& \text{let } k_c = \text{h}(y^{r \cdot c}) \text{ in} \\
& \text{let } cert = (\text{g}^c, \text{sign}(\text{g}^c, s)) \text{ in} \\
& \text{out}\bigl(ch, \{(r, cert)\}_{k_c}\bigr)
\end{aligned}$$

Since the card fails unlinkability even without a terminal being present, we can model a system involving only cards to simplify our analysis of the attack. The system in which the card can be used many times can be expressed as follows, where the public key $\text{pk}(s)$ of the payment system that manufactured the card (e.g., VISA, Mastercard, etc.) is made public by the attacker.

$$Sys_{BDH} \triangleq \text{new } s; \text{out}(out, \text{pk}(s)); !\text{new } c; !\text{new } ch; \text{out}(card, ch); C(s, c, ch)$$

As previously, the idealised specification can be obtained by removing the second replication, thereby insisting that each card is used only once.

$$Spec_{BDH} \triangleq \text{new } s; \text{out}(out, \text{pk}(s)); !\text{new } c, ch; \text{out}(card, ch); C(s, c, ch)$$

7.2 The strong privacy property: unlinkability

In what follows, we express what it means for the BDH key agreement protocol to fail unlinkability.

Theorem 7.3 $Sys_{BDH} \not\sim Spec_{BDH}$.

Due to the Hennessy-Milner property (Thm. 7.2), the above can be proven by providing a formula describing an attack. An attacker may perform the following steps when interacting with the system.

1. An attacker, posing as a malicious terminal, establishes a key with an honest card, then successfully decrypts the message z_2 and obtains the card's certificate $(g^c, \text{sign}(g^c, s))$.
2. Later the attacker poses as another malicious terminal to the card, runs a new session with the same card to obtain again the card's certificate $(g^c, \text{sign}(g^c, s))$, and hence recognises the card.

The above attack can be described by the following formula that is satisfied by the system Sys_{BDH}, but not by the specification $Spec_{BDH}$.

$$\begin{aligned}
&\langle \text{out}(out, pk_s) \rangle \\
&\langle \text{out}(card, c_1) \rangle \langle \text{out}(c_1, v_1) \rangle \langle \text{in}(c_1, g^{y_1}) \rangle \langle \text{out}(c_1, w_1) \rangle \\
&\langle \text{out}(card, c_2) \rangle \langle \text{out}(c_2, v_2) \rangle \langle \text{in}(c_2, g^{y_2}) \rangle \langle \text{out}(c_2, w_2) \rangle \\
&(\text{snd}(\text{dec}(w_1, \text{h}(v_1^{y_1}))) = \text{snd}(\text{dec}(w_2, \text{h}(v_2^{y_2}))))
\end{aligned}$$

The equation at the end of the attack corresponds exactly to the test of whether two certificates an attacker obtained from two sessions are equal. Below we explain this equation in more detail. It is possible for the system Sys_{BDH} to reach a state of the following form, where _ is some process:

$$A \triangleq \text{new } s, c, ch_1, ch_2, r_1, r_2;$$
$$\left[\left(\begin{cases} pk_s \mapsto \text{pk}(s), \\ u_1 \mapsto ch_1, \\ v_1 \mapsto (g^c)^{r_1}, \\ w_1 \mapsto \{(r_1, (g^c, \text{sign}(g^c, s)))\}_{\text{h}((g^{y_1})^{r_1 \cdot c})}, \\ u_2 \mapsto ch_2, \\ v_2 \mapsto (g^c)^{r_2}, \\ w_2 \mapsto \{(r_2, (g^c, \text{sign}(g^c, s)))\}_{\text{h}((g^{y_2})^{r_2 \cdot c})} \end{cases} \right) \mid \_ \right]$$

The idealised specification $Spec_{BDH}$, that models a system that is certainly unlinkable, reaches an extended process of the following form after performing the same actions.

$$B \triangleq \text{new } s, c_1, c_2, ch_1, ch_2, r_1, r_2;$$

$$\left[\left[\left\{\begin{array}{l} pk_s \mapsto \text{pk}(s), \\ u_1 \mapsto ch_1, \\ v_1 \mapsto (g^{c_1})^{r_1}, \\ w_1 \mapsto \{(r_1, (g^{c_1}, \text{sign}(g^{c_1}, s)))\}_{\text{h}((g^{y_1})^{r_1 \cdot c_1})}, \\ u_2 \mapsto ch_2, \\ v_2 \mapsto (g^{c_2})^{r_2}, \\ w_2 \mapsto \{(r_2, (g^{c_2}, \text{sign}(g^{c_2}, s)))\}_{\text{h}((g^{y_2})^{r_2 \cdot c_2})} \end{array}\right\}\right| - \right]\right]$$

Thus, since the actions are all matched, we have not yet violated the definition of a bisimulation by going through the transition steps. However, in this state, there is an equation that holds for the system that does not hold for the specification. That is, we have the following.

$$A \models \text{snd}(\text{dec}(w_1, \text{h}(v_1^{y_1}))) = \text{snd}(\text{dec}(w_2, \text{h}(v_2^{y_2})))$$

Yet the corresponding test fails for the specification as follows.

$$B \not\models \text{snd}(\text{dec}(w_1, \text{h}(v_1^{y_1}))) = \text{snd}(\text{dec}(w_2, \text{h}(v_2^{y_2})))$$

This explains the equation at the end of the test. It is comparing two messages that appear in different sessions with cards and testing whether they are equal. Ideally, they should not be if unlinkability holds, but that is not the case and hence this test can be used to trace cards.

7.2.4 Practice in verifying privacy

The message sequence chart for the Kim-Choi-Lee protocol [76] is presented in Fig. 7.4.

Exercise 7.1 Prove that the Kim-Choi-Lee protocol does not satisfy unlinkability by following these steps.

1. Model the roles of the protocol formally as threads.
2. Formulate the unlinkability of this protocol as an equivalence problem in terms of bisimilarity.
3. Provide an attack on unlinkability formally as a formula describing a trace.
4. Is the identity of the prover revealed if we make identities private names?
5. Is it possible to avoid the attack on unlinkability, perhaps by the prover sending an extra message to the verifier at the beginning in order to check recent aliveness of the prover? Provide a modified version of the protocol that is unlinkable.

Hint: try multiplying messages together when looking for attacks.

The following is an advanced exercise for reader wishing to explore further.

7.2 The strong privacy property: unlinkability

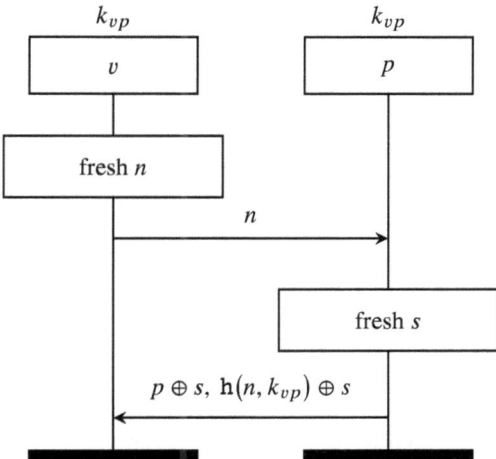

Fig. 7.4 The Kim-Choi-Lee protocol.

Exercise 7.2 Consider the following protocol modelling a private server r that only responds to a particular client i.

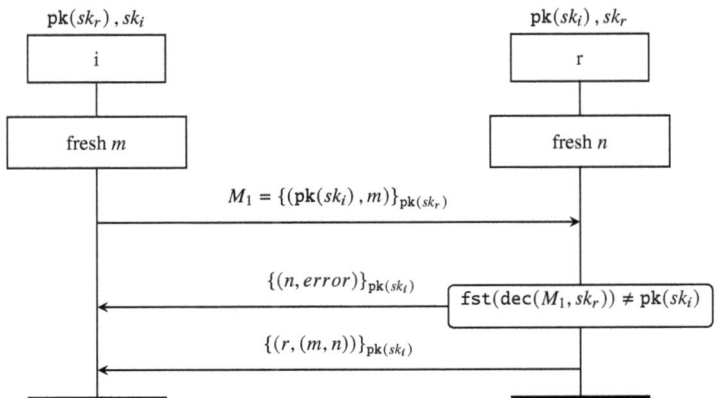

Express unlinkability of the above protocol. In both the system and specification, generate fresh private identities for each agent and output public keys in parallel to the execution of the client and server thread with those keys (this is so that the public key of one session is not associated with the public key of another session). In your model of the system, in addition, allow multiple sessions between a client and a server that are setup to talk to each other. In contrast, in the idealised specification assume that each client and server talk once and the identities involved are never reused. Argue whether or not unlinkability really holds with respect to your model.

There is an attack on unlinkability if we replace n with m in the message containing an error. Find and express the attack.

Discussion on privacy beyond protocols. Privacy problems are prevalent in systems and will require a cultural shift among developers to eradicate systematically. For a simple everyday privacy problem, consider the scenario where you hold some online account, say for social benefits. Suppose an attacker knows your ID, e.g., an email address serving as a user name, then they may be able to learn whether you hold an account simply via error messages. If an error message reveals that the password is incorrect only if the username exists, then, without authenticating, the attacker knows that the account exists. This a typical privacy oversight that requires conscious design, even if it is not strictly a protocol problem, the knowledge from this section is transferable. Suppose that the message when entering a wrong password does not compromise the privacy of the account holder, there may be further problems. For example, it may be that the message given in reply to a password recovery attempt, may reveal whether an account exists. For a concrete example, Mastodon does well here stating generically: "If your email address exists in our database, you will receive a password recovery link at your email address in a few minutes." The same message applies in all scenarios and only the authorised email account holder learns anything.

7.3 Putting it all together: a protocol combining unlinkability, agreement, distance bounding, and forward secrecy

This concluding section combines knowledge accumulated throughout the book to design a new protocol. By this point, we have studied quite a few properties of cryptographic protocols and have scrutinised a variety of examples that demonstrate a property (or the lack of) in isolation. We build a protocol that satisfies a combination of strong security and privacy properties we have encountered and elaborate on some that we have not yet explored. This protocol is a variant of the key agreement for card payments, similar to Blinded Diffie-Hellman in Fig. 2.9, while satisfying the following.

- An equivalence-based property of *unlinkability*.
- A time-based property of *Mafia-fraud resistance*.
- An authentication property of *injective agreement*.
- A confidentiality property of *forward secrecy* of the established key.

All properties have been defined previously in this book, except forward secrecy. Unlinkability means that, in this new protocol, it would be impossible for an active

7.3 Putting it all together   205

attacker (including a malicious terminal, that can engage the card in a session) to determine whether sessions are made with the same card. The distance-bounding property Mafia-fraud resistance ensures an honest terminal has guarantees that the card it communicates with is close. Injective agreement ensures that when a terminal believes it has authenticated the card, the card really ran the protocol and exchanged the same messages as the terminal. Forward secrecy enhances secrecy of the session key used to encrypted data after authentication by ensuring that an attacker cannot obtain the session key even if, in the future, the long-term keys are compromised.

7.3.1 Recapping agreement and distance bounding

Compared to the Blinded Diffie-Hellman protocol, shown to violate unlinkability in Sec. 7.2.3, we introduce two modifications. Firstly, we incorporate a fast phase in the style of Meadows' protocols (e.g., Fig. 6.8) to allow the terminal to check that the card is close. Secondly, we use Verheul signatures [110], a special type of digital signature that is homomorphic with respect to exponentiation, hence enabling the card to produce a new verifiable certificate in each session from the permanent certificate it holds.

We extend our message theory AE with new functions and respective equations.

$$
\begin{aligned}
\text{M, N} ::= &\ \ldots \\
&|\ \text{vpk(M)} &&\text{Verheul public key operation} \\
&|\ \text{vsign(M, N)} &&\text{Verheul signature} \\
&|\ \text{vcheck(M, N)} &&\text{Verheul signature verification}
\end{aligned}
$$

(SYNTAX OF MESSAGES EXTENDED WITH VERHEUL SIGNATURES)

$$\text{vcheck}(\text{vsign}(M, K), \text{vpk}(K)) =_{\text{AE}} M \qquad \text{vsign}(N, K)^M =_{\text{AE}} \text{vsign}(N^M, K)$$

(EQUATIONS FOR VERHEUL SIGNATURES)

The first equation above is standard, as seen before for signatures. The second equation above states that Verheul signatures are homomorphic with respect to exponentiation, that is, they allow the exponent to be pushed inside the signature. This allows us to blind (to raise to the same power) both the message and the signature, yet still verify this blinded pair using the $\text{vcheck}(\cdot, \cdot)$ function. These equations abstract away details of the underlying cryptography which use a device called bilinear pairings to check signatures on classes of pairing-friendly elliptic curves [20]. These operations are efficiently supported by dedicated circuitry on the current generation of smartcards.

FI am a bit confused about the fact that the two messages in the fast phase have no connection with each other but I did not spend much time on this.

Consider the proposal for an Unlinkable Distance-Bounding Blinded Diffie-Hellman protocol UDBB in Fig. 7.5. Similarly to BDH, the card holds its secret

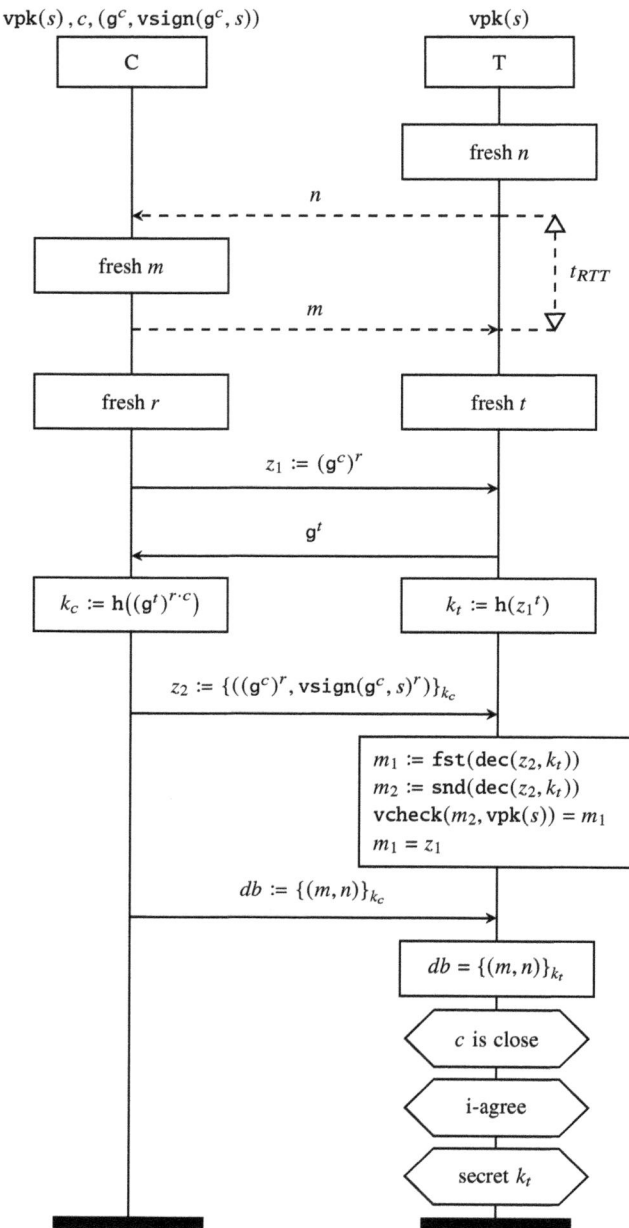

Fig. 7.5 Distance-bounding key agreement for card payments that satisfies unlinkability, injective agreement and secrecy of the session key.

7.3 Putting it all together

key c, and the Verheul certificate on its public key $(g^c, \mathsf{vsign}(g^c, s))$ issued by the payment system with the secret key s. The terminal holds only the Verheul public key of the payment system it supports $\mathsf{vpk}(s)$.

1. The protocol starts with the fast phase immediately. The terminal powers up the card sends its nonce n, receives the card's nonce m in reply and measures the round trip time.
2. The next two messages are identical to BDH from Sec. 7.2.3, i.e., the card sends $(g^c)^r$, its public key, blinded with a fresh scalar r. The terminal replies with the ephemeral public key g^t. At this point, both can locally derive the symmetric key $\mathsf{h}(g^{r \cdot t \cdot c})$.
3. Then the card uses the scalar r to blind the Verheul signature on its public key, pairs it with the key already blinded and sends this certificate encrypted to the terminal. The terminal then decrypts it, checks the signature and verifies that the blinded public key from the received certificate coincides with the first message.

Notice that the blinded certificate that the card sends appears different to the terminal in each session, hence the attack on unlinkability of BDH is avoided – it is not possible to recognise the card neither by its public key nor by the signature on it. Finally, the card sends encrypted the pair of nonces from the fast phase $\{(m, n)\}_{\mathsf{h}(g^{r \cdot t \cdot c})}$. The terminal checks if this pair is correct, which completes the authentication of the card by the terminal.

We next explain the ProVerif code, verifying Mafia-fraud resistance, injective agreement and the secrecy of the session key. As in previous chapters, the code explained here can be downloaded from Springer Nature Code Inside: https://github.com/sn-code-inside/security-protocols-and-threat-models.

In this context, the card is the prover and the terminal is the verifier. An equational theory for restricted scalar multiplication and approximating Verheul signature is provided below.

```
(* Scalar multiplication functions and equations *)
type point. (* elliptic curve public key *)
type scalar. (* elliptic curve secret key *)
const G: point [data]. (* DH group generator *)
fun smult(scalar, point): point. (* scalar multiplication *)

(* scalars acting on the DH group *)
equation forall a: scalar, b: scalar;
    smult(a, smult(b, G)) = smult(b, smult(a, G)).
equation forall a: scalar, b: scalar, c: scalar;
    smult(a, smult(b, smult(c, G))) = smult(b, smult(a, smult(c, G))).

(* Verheul signature *)
type svkey. (* Verheul signing key *)
type pvkey. (* Verheul public key *)
fun pkv(svkey): pvkey. (* Verheul public key operation *)
fun vsign(point, svkey): point. (* Verheul signature *)
```

(* checking Verheul signature, blinded with a scalar yields blinded message *)
reduc forall *pt*: point, *svk*: svkey;
   vcheck(vsign(*pt*, *svk*), pkv(*svk*)) = *pt*;
  forall a: scalar, *pt*: point, *svk*: svkey;
   vcheck(smult(a, vsign(*pt*, *svk*)), pkv(*svk*)) = smult(a, *pt*).

A distinction is made between points on an elliptic curve, which form a group, and scalar multiplication acting on that group. Notice that the generator of the group G is named in the equational theory for scalar multiplication meaning that only the first three scalar multiplications are subject to associativity and commutativity. It is important to explicitly acknowledge that ProVerif is working with respect to an approximation of the theory of scalar multiplication. Equations involving three scalars are considered due to the additional blinding factor in this protocol. The theory of Verheul signatures extends the theory of signatures such that it is possible to blind the signature with a scalar and then verify this blinded signature using the same public key.

Modelling symmetric encryption is straightforward – the same key is used to encrypt and decrypt messages.

type symkey. *(* key for symmetric encryption *)*
fun h(point): symkey. *(* deriving key from a DH group element operation *)*
fun senc(bitstring, symkey): bitstring. *(* symmetric encryption operation *)*
reduc forall *x*: bitstring, *y*: symkey; sdec(senc(*x*, *y*), *y*) = *x*. *(* decryption *)*

Mafia-fraud resistance is expressed as a query below, i.e., whenever the event close is triggered on the verifier's side, we check that the prover was present and, moreover, that it has received the nonce from the verifier during the fast phase.

event close(bitstring, bitstring).
event startff(bitstring).
event endff(bitstring).
event ffProver(bitstring).
query f: bitstring, p: bitstring, i, j, k: *time*;
  inj-event(close(f, p))$\wedge$
  inj-event(endff(f))@$k\wedge$
  inj-event(startff(f))@i
$\implies$
  (**inj-event**(ffProver(p))@$j \wedge i < j \wedge j < k$).

The query below expresses injective agreement as follows. An event marks the end of the protocol and is parameterised by the list of messages sent and received during the session. ProVerif then checks whether, at the moment when the verifier is ready to send its last message during the session, the verifier has the same list of messages and whether the respective verifier session uniquely corresponds to the prover's session.

7.3 Putting it all together

Notice that the uniqueness of this session correspondence is ensured by using the **inj–** prefix.

event PRunning(bitstring, point, point, bitstring, bitstring).
event VCommitWithP(bitstring, point, point, bitstring, bitstring).
query *nm*: bitstring, *bkey*: point, *ekey*: point, *ecert*: bitstring, *verif*: bitstring;
  inj-event(VCommitWithP(*nm*, *bkey*, *ekey*, *ecert*, *verif*))
  $\implies$
  inj-event(PRunning(*nm*, *bkey*, *ekey*, *ecert*, *verif*)).

To check the secrecy of the session key we use the standard query below.

event confidential(symkey).
query *k*: symkey; **event**(confidential(*k*)) $\wedge$ *attacker*(*k*).

Now we can specify the respective verifier and prover processes. Pay attention to where we put events to verify injective agreement. VCommitWithP is placed at the very end of the protocol, after the verifier has received the last message and ran all the required checks, whereas PRunning is placed right before the prover sends its final message.

let *Verifier*(ch: channel, *v*: bitstring, *p*: bitstring, *s*: svkey) =
  (* fast phase challenge *)
  new *f*: bitstring;
  event startff(*f*);
  new *n*: bitstring;
  out(ch, *n*);
  in(ch, *m*: bitstring);
  event endff(*f*);

  (* handshake *)
  new *t*: scalar;
  in(ch, *z1*: point);
  out(ch, smult(*t*, G));
  in(ch, *z2*: bitstring);
  let *k* = h(smult(*t*, *z1*)) **in**
  let (*cardKey*: point, *signature*: point) = sdec(*z2*, *k*) **in**
  if *cardKey* = vcheck(*signature*, pkv(*s*)) **then**
  if *cardKey* = *z1* **then**
  in(ch, *verif*: bitstring);

  (* check that nonces are correct *)
  if *verif* = senc((*m*, *n*), *k*) **then**
  event VCommitWithP((*n*, *m*), *z1*, smult(*t*, G), *z2*, *verif*);
  event close(*f*, *p*);
  event confidential(*k*).

let *Prover*(ch: channel, *p*: bitstring, *v*: bitstring, *c*: scalar, *s*: svkey) =
  (fast phase reponse *)*
  in(ch, *n*: bitstring);
  event ffProver(*p*);
  new *m*: bitstring;
  out(ch, *m*);

  (handshake *)*
  new *r*: scalar;
  let *spub*: point = smult(*r*, smult(*c*, G)) **in**
  out(ch, *spub*);
  in(ch, *y*: point);
  let *k* = h(smult(*r*, smult(*c*, *y*))) **in**
  let *cert* = (*spub*, smult(*r*, vsign(smult(*c*, G), *s*))) **in**
  out(ch, senc(*cert*, *k*));

  (prove that nonces are correct *)*
  event PRunning((*n*, *m*), *spub*, *y*, senc(*cert*, *k*), senc((*m*, *n*), *k*));
  out(ch, senc((*m*, *n*), *k*)).

Finally, the process specifying the system is as follows. The new system-wide private Verheul key is generated, and the respective verification Verheul key is made public. Then, we specify that there are infinitely many verifier sessions and, in parallel, infinitely many provers (cards) with the secret key c, each of which can run an infinite number of UDBB sessions.

free a, b: bitstring.
free ch: channel.
process new *s*: svkey; **out**(ch, pkv(*s*)); (
  (!*Verifier*(ch, b, a, *s*)) |
  (!**new** *c*: scalar; !*Prover*(ch, a, b, *c*, *s*)))

7.3.2 Unlinkability in ProVerif

To verify the unlinkability of the UDBB protocol, we follow the pattern explained at the beginning of this chapter and establish the equivalence between the system, where a prover can participate in many sessions, and the idealised world, where a prover can participate in at most one session. Since equivalence checking in ProVerif is incompatible with queries, we must use a separate file with all reachability queries

7.3 Putting it all together

removed. In this file, we replace the last part, where we specify the system, with the following code:

free a, b: bitstring.
free ch: channel.
equivalence
  new s: svkey; **out**(ch, pkv(s)); (
     (!*Verifier*(ch, b, a, s)) |
     (!**new** c: scalar; !*Prover*(ch, a, b, c, s)))
  new s: svkey; **out**(ch, pkv(s)); (
     (!*Verifier*(ch, b, a, s)) |
     (!**new** c: scalar; *Prover*(ch, a, b, c, s)))

At the time of writing, ProVerif runs forever on the above code. We can however use a trick to make ProVerif prove that UDBB indeed satisfies unlinkability. Since the prover shares no secret with the verifier, it can be simulated entirely by an attacker and hence we can drop the verifier from the above. This is an example of using compositional reasoning [70]. The code for this reduced equivalence problem is as follows.

free a, b: bitstring.
free ch: channel.
equivalence
  new s: svkey; **out**(ch, pkv(s));
    !**new** c: scalar; !*Prover*(ch, a, b, c, s)
  new s: svkey; **out**(ch, pkv(s));
    !**new** c: scalar; *Prover*(ch, a, b, c, s)

The above problem should terminate in ProVerif. In general, verifying equivalence is computationally more challenging for ProVerif than verifying queries—for a relatively short protocol like UDBB, it takes around 20 minutes on an Apple M3 chip with 16 GB of RAM, whereas for other properties, it is a matter of seconds.

7.3.3 Forward secrecy and beyond

We now come to the conclusion of our adventure, but wish to send you off with one more idea. Confidentiality and integrity are key notions in security, and are key to areas such as secure programming, not only protocols. They are properties that have associated with them a notion of flow from the past and into the future associated with them. If an object is designated as confidential then we expect that, into the future, it will remain confidential. If you receive an object then assurances on the integrity of its origin, in the past, increases trust.

Confidentiality and integrity can be expressed in many interesting ways, a space in which we foresee creativity, particularly as security and data protection regulation requires stronger guarantees on information systems. For example, non-repudiation protocols allow for a trusted third party to vouch for the integrity of a transaction *in the past*, despite adversarial action from some participants [78, 14]. An established notion of confidentiality for protocols is forward secrecy, that guarantees secrecy *into the future*, even if there is a data breach compromising keys [62].

ProVerif has features allowing comprehensive formulations of forward secrecy to be expressed, so we go straight to discussing tooling. We continue the example from the previous section, by make two changes. Firstly, we modify the network configuration and annotate it with **phase** labels as follows.

free a, b: bitstring.
free ch: channel.
process new s: svkey; **out**(ch, pkv(s));
   (
     (!*Verifier*(ch, b, a, s)) |
     (
       !**new** c: scalar; (
         (!*Prover*(ch, a, b, c, s)) |
         phase 1; *(* Provers also run post compromise using same identity c *)*
         (!*Prover*(ch, a, b, c, s)))
) |
     phase 1 ; *(* the secret key becomes compromised *)*
     out(ch , s)
)

The phases are used to indicate what happens before and after the compromise of a long-term secret. Before the long term secret s is compromised, there are sessions with honest provers and verifiers, with provers are permitted to reuse identities, generated by **new** c in the code above. After the compromise, sequentially after **phase** markers, you can see that the following things happen.

- The long-term key s is output, making it available to the attacker.
- The prover can still execute after the phase marker and is allowed to continue to reuse their identities c created in the initial phase.

We deliberately only model the verifier in the first phase and not the second phase. The verifier process can be removed from the second phase, by compositional reasoning. Specifically, all actions of the verifier can be simulated by a Dolev-Yao attacker, and, furthermore, there should be no confidentiality claims for such executions, since we know that sessions after a compromise do not preserve confidentiality. In contrast, the verifier does appear explicitly in the initial phase, since we are interested in the confidentiality of session keys generated by the verifier during sessions before the long-term key is compromised.

To model the fact that we are interested in whether the session keys generated by the verifier in the initial phase remain confidential during the second phase (post

7.3 Putting it all together 213

compromise), we modify the end of the verifier processes from the previous section as follows.

phase 1; *(* check confidentiality post compromise *)*
event confidential(k).

Observe this means that verifier sessions can be executed during the initial phase of the protocol, but claims about the confidentiality of the session key generated are only checked during the second phase of the protocol. Thus we have modelled forward secrecy with respect to a compromise where the long-term secret key common to all verifiers is leaked.

7.3.4 Practice in using tools on larger case studies

The final exercises of this book push the reader to examine further the protocol developed in this section. Questions encompass all properties studied in this book. A separate exercise on the PACE protocol provides further practice with verifying forward secrecy and unlinkability.

Exercise 7.3 Attempt the following tasks for practice based on the substantial case study in this section.

1. Assemble the code for UDBB and check that all security and privacy properties described hold. The code should be assembled in separate files for unlinkability and forward secrecy.
2. Strengthen the threat model for forward secrecy such that also all identities of provers, c, may be compromised in the second phase. Why, by reasoning with respect to compositionality, can the prover processes post compromise be removed entirely from the model, given that s and c are leaked for each prover?
3. Check also recent aliveness from the perspective of the terminal for the UDBB protocol.
4. Consider the UDBB protocol in Fig. 7.5, but without the check $m_1 = z_1$ by the verifier at the end of the handshake. Comment out the respective line in the verifier process in both files. Draw an MSC for any resulting attacks on the security properties checked.
5. Notice that authentication and secrecy are not mutual in this protocol. Find an attack on secrecy of the session key from the perspective of the prover.
6. Modify the protocol so that the prover also authenticates the card. Check injective agreement for your protocol from the perspective of the prover.
7. Model and verify forward secrecy from the perspective of the prover for your new protocol. When doing so, ensure that you create two versions of the prover process such that (1) the confidentiality of keys generated by provers running in the initial phase is only checked in the second phase, while (2) confidentiality is not checked at all for provers running in the second phase.

8. Under what assumption does distance hijacking hold for this protocol?

The UDBB protocol investigated above combines two protocols in the literature proposed for enhancing the security and privacy of card and phone payments: the UBDH protocol [69] and a proposal for adding distance bounding to the ISO standard layers for wireless communication [99]. The former protocol has been extended to a multiparty protocol, named UTX, involving the full generation of application cryptograms and a round trip between the terminal and the payment system [38]. In the UTX protocol, the bank can authenticate the card and terminal, while the terminal can authenticate the bank and card. ProVerif code verifying authentication for that larger protocol is available in a repository referenced from the paper. Checking that authentication properties hold for UTX can be verified using conventional hardware. On the other hand, unlinkability is verified on paper, since protocols of this form remain out of scope of tools at the time of writing.

The following protocol is provided for extra practice. It ties together the beginning and end of this book, since we mentioned forward secrecy of the PACE ePassport protocol in Sec. 2.2.11.

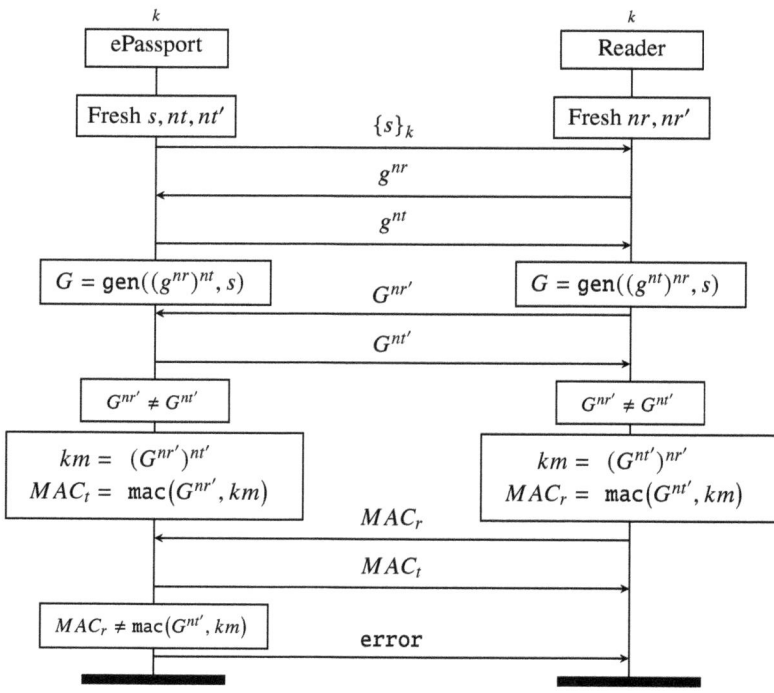

Fig. 7.6 The PACE ePassport protocol.

7.3 Putting it all together

Exercise 7.4 Consider now the PACE ePassport protocol in Fig. 7.6.

1. The PACE protocol satisfies forward secrecy, while the BAC protocol does not. Model forward secrecy for BAC and PACE to confirm this.
2. Is there an attack on unlinkability for the PACE protocol? You may use any tool or even the literature to find one.
3. Check whether agreement fails if the inequality checks in Fig. 7.6 are omitted?

Send off. This book aims to bring the reader to the edge of the state-of-the-art of symbolic verification of protocols. So, what more can we do? The literature on formally verifying security protocols is vast and spans decades, so there are many more examples of protocols and properties out there. Here we mention a few horizons for research besides applying the methods mastered to more protocols and threats. In this book, we have shown how to use tools for verifying symbolic protocols. However, these tools are software, which can contain bugs themselves. As fairly mature tools, they have been thoroughly tested but verifying their correctness would yield even higher assurance in the case that no flaws have been reported. This would be a huge effort and an alternative is checking their output: e.g., inductive invariants as certificates of correctness could be checked independently. Similarly, the meta theory of the applied π-calculus could be mechanically proven using proof assistants. Such independent checking improves trust in tool chains, although mistakes when modelling are more likely to be a concern. To reduce such mistakes, more automation could be introduced, such as tooling for precise correspondences between threads representing roles and MSCs. In addition, the larger case studies in this book already showcased some limits that tools can hit. Hence, tools can always be improved as we drive towards handling still larger protocols, richer threat models and more complex message theories. The reader, by now, is however equipped to apply threat modelling in practice in order to inform security engineering decisions.

References

1. EMV Integrated Circuit Card Specifications for Payment Systems. Book 1: Application Independent ICC to Terminal Interface Requirements. Tech. rep., EMVCo LLC (2011). URL https://www.emvco.com/document-search/. Accessed: 26-08-2021
2. EMV Integrated Circuit Card Specifications for Payment Systems. Book 2: Security and Key Management. Tech. rep., EMVCo LLC (2011). URL https://www.emvco.com/document-search/. Accessed: 26-08-2021
3. EMV Integrated Circuit Card Specifications for Payment Systems. Book 3: Application Specification. Tech. rep., EMVCo LLC (2011). URL https://www.emvco.com/document-search/. Accessed: 26-08-2021
4. EMV Integrated Circuit Card Specifications for Payment Systems. Book 4: Cardholder, Attendant, and Acquirer Interface Requirements. Tech. rep., EMVCo LLC (2011). URL https://www.emvco.com/document-search/. Accessed: 26-08-2021
5. EMV ECC key establishment protocols. RFC until 28th January 2013, EMVCo LLC (2012). URL http://www.emvco.com/specifications.aspx?id=243. Accessed: 01-04-2020
6. Common criteria for information technology Security evaluation. Part 2: Security functional components. Tech. rep. (2017). Version 3.1, Revision 5, CCMB-2017-04-002
7. Transaction Processing Rules. Tech. rep., Mastercard (2024). URL https://www.mastercard.us/content/dam/public/mastercardcom/na/global-site/documents/transaction-processing-rules.pdf. Accessed: 12-12-2024
8. Abadi, M., Blanchet, B., Fournet, C.: The applied pi calculus: Mobile values, new names, and secure communication. J. ACM **65**(1), 1:1–1:41 (2018)
9. Abadi, M., Rogaway, P.: Reconciling two views of cryptography (the computational soundness of formal encryption). Journal of Cryptology **15**(2), 103–127 (2002)
10. Ahn, K.Y., Horne, R., Tiu, A.: A characterisation of open bisimilarity using an intuitionistic modal logic. Log. Methods Comput. Sci. **17**(3) (2021)
11. Ali, M.A.: Does the online card payment system unwittingly facilitate fraud? Ph.D. thesis, Newcastle University (2018)
12. Anderson, R.: Security Engineering: a Guide to Building Dependable Distributed Systems – Third Edition. Wiley (2020)
13. Arapinis, M., Chothia, T., Ritter, E., Ryan, M.: Analysing unlinkability and anonymity using the applied pi calculus. In: 23rd IEEE Computer Security Foundations Symposium, pp. 107–121 (2010)
14. Asokan, N., Shoup, V., Waidner, M.: Asynchronous protocols for optimistic fair exchange. In: Proceedings. 1998 IEEE Symposium on Security and Privacy (Cat. No. 98CB36186), pp. 86–99. IEEE (1998)
15. Aubert, C., Horne, R., Johansen, C.: Diamonds for security: A non-interleaving operational semantics for the applied pi-calculus. In: Proc. 33rd Int. Conf. on Concurrency Theory

(CONCUR'22), *LIPIcs*, vol. 243, pp. 30:1–30:26. Schloss Dagstuhl - Leibniz-Zentrum für Informatik (2022)
16. Avoine, G., Bingöl, M.A., Boureanu, I., Čapkun, S., Hancke, G., Kardaş, S., Kim, C.H., Lauradoux, C., Martin, B., Munilla, J., et al.: Security of distance-bounding: A survey. ACM Computing Surveys (CSUR) **51**(5), 94:1–33 (2018). DOI 10.1145/3264628
17. Avoine, G., Bultel, X., Gambs, S., Gérault, D., Lafourcade, P., Onete, C., Robert, J.M.: A terrorist-fraud resistant and extractor-free anonymous distance-bounding protocol. In: Proc. 2017 ACM Asia Conference on Computer and Communications Security (ASIACCS'17), pp. 800–814. ACM, New York, USA (2017)
18. Baader, F., Nipkow, T.: Term rewriting and all that. Cambridge University Press, USA (1998)
19. Backus, J.W.: The syntax and semantics of the proposed international algebraic language of the Zurich ACM-GAMM conference. Proc. Int. Conf. on Information Processing (1959)
20. Barreto, P.S., Naehrig, M.: Pairing-friendly elliptic curves of prime order. In: Proc. Int. workshop on selected areas in cryptography, pp. 319–331. Springer (2005)
21. Basin, D., Cremers, C.: Modeling and analyzing security in the presence of compromising adversaries. In: Proc. 15th European Symposium On Research In Computer Security (ESORICS'10), *LNCS*, vol. 6345, pp. 340–356. Springer (2010)
22. Basin, D., Cremers, C., Dreier, J., Sasse, R.: Modeling and Analyzing Security Protocols with Tamarin: A Comprehensive Guide. Information Security and Cryptography. Springer (2025)
23. Basin, D., Dreier, J., Hirschi, L., Radomirovic, S., Sasse, R., Stettler, V.: A formal analysis of 5G authentication. In: Proc. 25th ACM Conf. on Computer and Communications Security (CCS'18), pp. 1383–1396. ACM (2018)
24. Basin, D., Sasse, R., Toro-Pozo, J.: Card brand mixup attack: Bypassing the PIN in non-Visa cards by using them for Visa transactions. In: Proc. 30th USENIX Security Symposium (USENIX Security'21), pp. 179–194. USENIX Association (2021)
25. Basin, D., Sasse, R., Toro-Pozo, J.: The EMV standard: Break, fix, verify. In: Proc. IEEE Symp. on Security and Privacy (S&P'21), pp. 1766–1781. IEEE (2021)
26. Bender, J., Fischlin, M., Kügler, D.: Security analysis of the PACE key-agreement protocol. In: Proc. 12th Int. Conf. on Information Security (ISC'09), *LNCS*, vol. 5735, pp. 33–48. Springer (2009)
27. Bisht, R.K., Dhami, H.S.: Discrete Mathematics. Oxford University Press (2015)
28. Blanchet, B.: Modeling and verifying security protocols with the applied pi calculus and ProVerif. Foundations and Trends in Privacy and Security **1**(1-2), 1–135 (2016)
29. Blanchet, B., Cheval, V., Cortier, V.: ProVerif with lemmas, induction, fast subsumption, and much more. In: 2022 IEEE Symp. on Security and Privacy (SP), pp. 69–86. IEEE (2022)
30. Blanchet, B., Podelski, A.: Verification of cryptographic protocols: Tagging enforces termination. In: Proc. 6th Int. Conf. on Foundations of Software Science and Computation Structures (FoSSaCS'03), *LNCS*, vol. 2620, pp. 136–152. Springer (2003)
31. Blanchet, B., Smyth, B., Cheval, V., Sylvestre, M.: ProVerif 2.05: Automatic cryptographic protocol verifier, user manual and tutorial (2023). URL https://bblanche.gitlabpages.inria.fr/proverif/manual.pdf. Accessed: 19-02-2025
32. Bond, M., Choudary, M.O., Murdoch, S.J., Skorobogatov, S.P., Anderson, R.J.: Be prepared: The EMV preplay attack. IEEE Secur. Priv. **13**(2), 56–64 (2015)
33. Bond, M., Choudary, O., Murdoch, S.J., Skorobogatov, S., Anderson, R.: Chip and skim: Cloning EMV cards with the pre-play attack. In: Proc. IEEE Symp. on Security and Privacy (S&P'14), pp. 49–64. IEEE (2014)
34. Boreale, M., Buscemi, M.G.: A method for symbolic analysis of security protocols. Theor. Comput. Sci. **338**(1-3), 393–425 (2005)
35. Boureanu, I., Chothia, T., Debant, A., Delaune, S.: Security analysis and implementation of relay-resistant contactless payments. In: Proc. ACM Conf. on Computer and Communications Security (CCS'20), pp. 879–898. ACM (2020)
36. Braun, C.H., Horne, R., Käfer, T., Mauw, S.: SSI, from specifications to protocol? Formally verify security! In: Proc. ACM Web Conference 2024, (WWW'24), pp. 1620–1631. ACM (2024)

References

37. van den Breekel, J., Ortiz-Yepes, D.A., Poll, E., de Ruiter, J.: EMV in a nutshell. Tech. rep., Radboud University (2016)
38. Bursuc, S., Horne, R., Mauw, S., Yurkov, S.: Provably unlinkable smart card-based payments. In: Proc. 30th ACM Conference on Computer and Communications Security (CCS'23), pp. 1392–1406. ACM (2023)
39. Canetti, R.: Universally composable security: A new paradigm for cryptographic protocols. In: Proceedings 42nd IEEE Symposium on Foundations of Computer Science, pp. 136–145. IEEE (2001)
40. Cheval, V., Jacomme, C., Kremer, S., Künnemann, R.: SAPIC+: Protocol verifiers of the world, unite! In: Proc. 31st USENIX Security Symposium (USENIX Security'22), pp. 3935–3952. USENIX Association (2022)
41. Cheval, V., Kremer, S., Rakotonirina, I.: DEEPSEC: Deciding equivalence properties in security protocols. Theory and practice. In: Proc. IEEE Symp. on Security and Privacy (S&P'18), pp. 529–546. IEEE (2018)
42. Chothia, T., Clee, A., Boureanu, I., Pavlides, G.: More is less: Extra features in contactless payments break security. In: Proc. 34th USENIX Security Symposium (USENIX Security'25). USENIX Association (2025). In press
43. Chothia, T., de Ruiter, J., Smyth, B.: Modelling and analysis of a hierarchy of distance bounding attacks. In: Proc. 27th USENIX Security Symposium (USENIX Security'18), pp. 1563–1580. USENIX Association (2018)
44. Coron, J., Gouget, A., Icart, T., Paillier, P.: Supplemental access control (PACE v2): Security analysis of PACE integrated mapping. In: D. Naccache (ed.) Cryptography and Security: From Theory to Applications - Essays Dedicated to Jean-Jacques Quisquater on the Occasion of His 65th Birthday, *LNCS*, vol. 6805, pp. 207–232. Springer (2012)
45. Cortier, V., Kremer, S.: Formal models and techniques for analyzing security protocols: A tutorial. Foundations and Trends in Programming Languages **1**(3), 151–267 (2014)
46. Cremers, C., Mauw, S.: Operational Semantics and Verification of Security Protocols. Information Security and Cryptography. Springer (2012)
47. Cremers, C.J.F.: The Scyther tool: Verification, falsification, and analysis of security protocols. In: Proc. 20th Int. Conf. on Computer Aided Verification (CAV'08), *LNCS*, vol. 5123, pp. 414–418. Springer (2008)
48. Dolev, D., Yao, A.: On the security of public-key protocols. IEEE Transactions on Information Theory **29**(2), 198—208 (1983)
49. Drimer, S., Murdoch, S.J.: Keep your enemies close: Distance bounding against smartcard relay attacks. In: Proc. 16th USENIX Security Symposium (USENIX Security'07), pp. 87–102. USENIX Association (2007)
50. Emms, M., Arief, B., Freitas, L., Hannon, J., van Moorsel, A.: Harvesting high value foreign currency transactions from EMV contactless credit cards without the PIN. In: Proc. 21st ACM Conf. on Computer and Communications Security (CCS'14), p. 716–726. ACM (2014)
51. EMVCo: Worldwide EMV Deployment Statistics (2022). URL https://www.emvco.com/about-us/worldwide-emv-deployment-statistics/. Accessed: 26-01-2023
52. EMVCo: EMV contactless specifications for payment systems, Book C-8, Kernel 8 specification. Tech. Rep. Version 1.1 (2023)
53. Engelhardt, M., Pfeiffer, F., Finkenzeller, K., Biebl, E.: Extending ISO/IEC 14443 type A eavesdropping range using higher harmonics. In: Proc. European Conf. on Smart Objects, Systems and Technologies (SmartSysTech'13), pp. 1–8. IEEE (2013)
54. Escobar, S., Meadows, C., Meseguer, J.: Maude-npa: Cryptographic protocol analysis modulo equational properties. In: A. Aldini, G. Barthe, R. Gorrieri (eds.) Foundations of Security Analysis and Design V: FOSAD 2007/2008/2009 Tutorial Lectures, pp. 1–50. Springer (2009)
55. Esposito, C., Horne, R., Robaldo, L., Buelens, B., Goesaert, E.: Assessing the Solid protocol in relation to security and privacy obligations. Information **14**(7), 411 (2023)
56. Feldhofer, M., Dominikus, S., Wolkerstorfer, J.: Strong authentication for RFID systems using the AES algorithm. In: Proc. 6th Int. Workshop on Cryptographic Hardware and Embedded Systems (CHES'04), *LNCS*, vol. 3156, pp. 357–370. Springer Berlin Heidelberg (2004)

57. Fett, D., Campbell, B., Bradley, J., Lodderstedt, T., Jones, M., Waite, D.: OAuth 2.0 Demonstrating Proof of Possession (DPoP). Tech. Rep. RFC 9449, Internet Engineering Task Force (IETF) (2023)
58. Fett, D., Küsters, R., Schmitz, G.: A comprehensive formal security analysis of OAuth 2.0. In: Proc. 23rd ACM Conf. on Computer and Communications Security (CCS'16), pp. 1204–1215. ACM (2016)
59. Filimonov, I., Horne, R., Mauw, S., Smith, Z.: Breaking unlinkability of the ICAO 9303 standard for e-passports using bisimilarity. In: Proc. 24th European Symposium On Research In Computer Security (ESORICS'19), Part 1, *LNCS*, vol. 11735, pp. 577–594. Springer (2019)
60. Gallier, J.: Discrete Mathematics. Universitext. Springer (2011)
61. Gentzen, G.: Investigations into logical deductions. In: M.E. Szabo (ed.) The Collected Papers of Gerhard Gentzen, pp. 68–131. North-Holland Publishing Co., Amsterdam (1969)
62. Günther, C.G.: An identity-based key-exchange protocol. In: J.J. Quisquater, J. Vandewalle (eds.) Advances in Cryptology — EUROCRYPT '89, pp. 29–37. Springer Berlin Heidelberg, Berlin, Heidelberg (1990)
63. Habraken, R., Dolron, P., Poll, E., de Ruiter, J.: An RFID skimming gate using higher harmonics. In: Proc. 11th Int. Workshop on Radio Frequency Identification (RFIDSec'15), pp. 122–137. Springer (2015)
64. Hathhorn, C., Rosu, G.: Dealing with C's original sin. IEEE Software **36**(5), 24–28 (2019)
65. Hennessy, M., Milner, R.: Algebraic laws for nondeterminism and concurrency. J. ACM **32**(1), 137–161 (1985)
66. Honda, K., Yoshida, N., Carbone, M.: Multiparty asynchronous session types. In: Proc. 35th ACM SIGPLAN-SIGACT Symposium on Principles of Programming Languages (POPL'08), pp. 273–284. ACM (2008)
67. Hopcroft, J.E., Motwani, R., Ullman, J.D.: Introduction to automata theory, languages, and computation, 3rd Edition. Pearson international edition. Addison-Wesley (2007)
68. Horne, R., Mauw, S.: Discovering epassport vulnerabilities using bisimilarity. Log. Methods Comput. Sci. **17**(2), 24 (2021)
69. Horne, R., Mauw, S., Yurkov, S.: Unlinkability of an improved key agreement protocol for EMV 2nd gen payments. In: S. Calzavara (ed.) 35th IEEE Computer Security Foundations Symposium, CSF 2022, Haifa, Israel, August 7-10, 2022, pp. 364–379. IEEE (2022). DOI 10.1109/CSF54842.2022.9919666
70. Horne, R., Mauw, S., Yurkov, S.: When privacy fails, a formula describes an attack: A complete and compositional verification method for the applied π-calculus. Theor. Comput. Sci. **959**, 113842 (2023)
71. ICAO: Machine readable travel documents. Part 11: Security mechanisms for MRTDs. (2015). URL https://www.icao.int/publications/Documents/9303_p11_cons_en.pdf. Accessed: 21-07-2025
72. Igarashi, A., Pierce, B.C., Wadler, P.: Featherweight Java: a minimal core calculus for Java and GJ. ACM Transactions on Programming Languages and Systems (TOPLAS) **23**(3), 396–450 (2001)
73. Jones, S.P.: Haskell 98 language and libraries: The revised report. Cambridge University Press (2003)
74. Kahn, G.: The semantics of a simple language for parallel programming. In: Proc. 6th IFIP Congress (Information Processing'74), pp. 471–475. North-Holland Publishing Co. (1974)
75. Katz, J., Lindell, Y.: Introduction to modern cryptography: principles and protocols. Chapman and hall/CRC (2007)
76. Kim, I.J., Choi, E.Y., Lee, D.H.: Secure mobile RFID system against privacy and security problems. In: Proc. 3rd Int. Workshop on Security, Privacy and Trust in Pervasive and Ubiquitous Computing (SecPerU'07), pp. 67–72. IEEE (2007)
77. Klein, G., Elphinstone, K., Heiser, G., Andronick, J., Cock, D., Derrin, P., Elkaduwe, D., Engelhardt, K., Kolanski, R., Norrish, M., et al.: seL4: Formal verification of an OS kernel. In: Proc. 22nd ACM SIGOPS symposium on Operating systems principles (SOSP'09), pp. 207–220. ACM (2009)

78. Kremer, S., Markowitch, O., Zhou, J.: An intensive survey of fair non-repudiation protocols. Computer communications **25**(17), 1606–1621 (2002)
79. Kumar, R., Myreen, M.O., Norrish, M., Owens, S.: CakeML: a verified implementation of ML. ACM SIGPLAN Notices **49**(1), 179–191 (2014)
80. Liu, Y., Kasper, T., Lemke-Rust, K., Paar, C.: E-passport: Cracking basic access control keys. In: Proc. On the Move to Meaningful Internet Systems 2007: CoopIS, DOA, ODBASE, GADA, and IS (OTM'07), Part II, *LNCS*, vol. 4804, pp. 1531–1547. Springer (2007)
81. Lowe, G.: Breaking and fixing the Needham-Schroeder public-key protocol using FDR. Softw. Concepts Tools **17**(3), 93–102 (1996)
82. Lowe, G.: A hierarchy of authentication specifications. In: Proc. 10th Computer Security Foundations Workshop (CSFW'97), pp. 31–43. IEEE (1997)
83. Luu, L., Chu, D.H., Olickel, H., Saxena, P., Hobor, A.: Making smart contracts smarter. In: Proc. 23rd ACM Conf. on Computer and Communications Security (CCS'16), pp. 254–269. ACM (2016)
84. Mauw, S., Reniers, M.A.: Operational semantics for MSC'96. Computer Networks **31**(17), 1785–1799 (1999)
85. Mauw, S., Smith, Z., Toro-Pozo, J., Trujillo-Rasua, R.: Distance-bounding protocols: Verification without time and location. In: Proc. IEEE Symp. on Security and Privacy (S&P'18), pp. 549–566. IEEE (2018)
86. Meadows, C., Poovendran, R., Pavlovic, D., Chang, L., Syverson, P.: Distance bounding protocols: Authentication logic analysis and collusion attacks. In: Secure Localization and Time Synchronization for Wireless Sensor and Ad Hoc Networks, *Advances in Information Security*, vol. 30, pp. 279–298. Springer US (2007)
87. Meier, S., Schmidt, B., Cremers, C., Basin, D.: The TAMARIN prover for the symbolic analysis of security protocols. In: Proc. 25th Int. Conf. Computer Aided Verification (CAV'13), *LNCS*, vol. 8044, pp. 696–701. Springer (2013)
88. Menezes, A., van Oorschot, P.C., Vanstone, S.A.: Handbook of Applied Cryptography (1st ed.). CRC Press (1996)
89. Milner, R.: A theory of type polymorphism in programming. J. Comput. Syst. Sci. **17**(3), 348–375 (1978)
90. Milner, R.: A Calculus of Communicating Systems, *LNCS*, vol. 92. Springer (1980)
91. Milner, R.: Communicating and mobile systems - the Pi-calculus. Cambridge University Press (1999)
92. Milner, R., Parrow, J., Walker, D.: A calculus of mobile processes, Part I and II. Information and Computation **100**(1), 1–100 (1992)
93. Morgan, J., Coburn, A., Bosquet, M.: Solid-OIDC primer. version 0.1.0. URL https://solidproject.org/TR/oidc-primer. Accessed: 21-07-2025
94. Murdoch, S.J., Drimer, S., Anderson, R., Bond, M.: Chip and PIN is broken. In: Proc. IEEE Symp. on Security and Privacy (S&P'10), pp. 433–446. IEEE (2010)
95. Needham, R.M., Schroeder, M.D.: Using encryption for authentication in large networks of computers. Commun. ACM **21**(12), 993–999 (1978)
96. Pancho, S.: Paradigm shifts in protocol analysis. In: Proc. Workshop on New Security Paradigms (NSPW'99), pp. 70–79. ACM (1999)
97. Peltz, C.: Web services orchestration and choreography. Computer **36**(10), 46–52 (2003)
98. Plotkin, G.D.: The origins of structural operational semantics. The Journal of Logic and Algebraic Programming **60-61**, 3–15 (2004)
99. Radu, A.I., Chothia, T., Newton, C.J., Boureanu, I., Chen, L.: Practical EMV relay protection. In: Proc. IEEE Symp. on Security and Privacy (S&P'22), pp. 1737–1756. IEEE (2022)
100. Reisig, W.: Petri nets: An introduction, vol. 4. Springer Science & Business Media (2012)
101. Ryan, P.Y., Schneider, S., Goldsmith, M., Lowe, G., Roscoe, B.: The Modelling and Analysis of Security Protocols: The CSP Approach, first edn. Addison-Wesley Professional (2000)
102. Sakimura, N., Bradley, J., Agarwal, N.: Proof Key for Code Exchange by OAuth Public Clients. Tech. Rep. RFC 7636, Internet Engineering Task Force (IETF) (2015)
103. Sangiorgi, D., Walker, D.: The Pi-Calculus - a theory of mobile processes. Cambridge University Press (2001)

104. Schaller, P., Schmidt, B., Basin, D.A., Capkun, S.: Modeling and verifying physical properties of security protocols for wireless networks. In: Proc. 22nd IEEE Computer Security Foundations Symposium, (CSF'09), pp. 109–123. IEEE (2009)
105. zu Selhausen, K.M., Fett, D.: OAuth 2.0 Authorization Server Issuer Identification. Tech. Rep. RFC 9207, Internet Engineering Task Force (IETF) (2022)
106. Tidy, J.: Crypto hacker offered reward after $600m heist. Cyber security reporter, BBC News (2021). URL https://www.bbc.com/news/business-58193396. Accessed: 13-08-2021
107. Tiu, A.F., Gore, R., Dawson, J.: A Proof Theoretic Analysis of Intruder Theories. Logical Methods in Computer Science **6**(3) (2010)
108. Ungoed-Thomas, J.: Revealed: car industry was warned keyless vehicles vulnerable to theft a decade ago. The Guardian (2024). URL https://www.theguardian.com/uk-news/2024/feb/24/revealed-car-industry-was-warned-keyless-vehicles-vulnerable-to-theft-a-decade-ago. Accessed: 24-04-2024
109. Urban, C., Pitts, A.M., Gabbay, M.J.: Nominal unification. Theoretical Computer Science **323**(1-3), 473–497 (2004)
110. Verheul, E.R.: Self-blindable credential certificates from the Weil pairing. In: Proc. 7th Int. Conf. on the Theory and Application of Cryptology and Information Security (ASIACRYPT'01), *LNCS*, vol. 2248, pp. 533–551. Springer (2001)

Index

abstract syntax, 48
accessibility, 106
active substitution, 63
agreement, 128
application cryptogram, 33
applied π-calculus, 51
attack, 75

BAC protocol, 23, 193
bisimilarity, 195
Blinded Diffie-Hellman protocol, 199

classical $\mathcal{FM}$, 198
co-induction, 76
Common Criteria, 191
compositionality, 142, 211, 212
constraint system, 105
context-free grammar, 49
correspondence assertion, 134
Cross-Site Request Forgery, 142

derivation tree, 65
distance hijacking, 177
Dolev-Yao threat model, 11, 97
DPoP token, 145

EMV protocol, 4, 28, 168, 199
ePassports, 9
equality, 129
extended process, 63
extrude, 68
extrusion, 133

fast phase, 170
Feldhofer protocol, 23
forward secrecy, 28, 212
free variables, 53, 66

fresh, 54

Hennessy-Milner property, 198
HTTP protocol, 143

idempotent substitution, 56, 105
inequalities, 105
inequality, 122, 129
inference rule, 64
injective agreement, 4, 14, 129, 139, 208
intuitionistic logic, 106

labelled transition, 65
law of the excluded middle, 106
lazy trace, 123

Mafia fraud, 171, 208
Meadows 1 protocol, 169
Meadows 2 protocol, 178
Message Authentication Code, 25
message sequence chart, 12, 57
messages, 49
mixup attack, 145
mobility, 134
modal logic, 74, 198
monotonicity, 101
most general unifiers, 122
multiparty authentication, 151
multiparty protocols, 142
mutual authentication, 16

Needham-Schroeder protocol, 89
Needham-Schroeder-Lowe protocol, 118
networks, 79
normal form, 64

Offline Data Authentication, 34

OpenID Connect protocol, 143
Optical Character Recognition, 10

PACE protocol, 214
PIN bypass attack, 41
prefix, 65
ProVerif, 20, 78, 134, 207
public key, 31, 50

recent aliveness, 165
reflection attack, 18
RFC, 142
RFC 7636, 150
RFC 9207, 150

Scyther, 19, 96
secrecy, 74
signature, 31, 131

single-sign-on protocol, 142
solution, 105
sub-formula property, 101
subsequence, 128
substitutions, 52
symmetric keys, 141

tagging, 149
threads, 50
TLS protocol, 143
TREAD protocol, 171

Verheul signature, 205

weak aliveness, 162

XOR, 167

GPSR Compliance

The European Union's (EU) General Product Safety Regulation (GPSR) is a set of rules that requires consumer products to be safe and our obligations to ensure this.

If you have any concerns about our products, you can contact us on

ProductSafety@springernature.com

In case Publisher is established outside the EU, the EU authorized representative is:

Springer Nature Customer Service Center GmbH
Europaplatz 3
69115 Heidelberg, Germany

www.ingramcontent.com/pod-product-compliance
Ingram Content Group UK Ltd.
Pitfield, Milton Keynes, MK11 3LW, UK
UKHW022203230426
470311UK00001BA/12